The Rocks of Georgia

by

Dan D. Williams

Dan Williams has been a Forest Manager, Environmental Educator and Writer with the Warnell School of Forestry and Natural Resources, University of Georgia for 30 years

For additional copies of this book, please email Dan:

williams@warnell.uga.edu
or visit the web site:
http://possumpublications.com/

Table of Contents

CHAPTER 1. GEORGIA'S PHYSIOGRAPHIC AND GEOLOGIC PROVINCES

Most folks know Georgia is divided into five physiographic provinces based mainly on topography (land shape). The color coded map below (a Google Maps screen shot) shows these provinces. To make words on the map more readable, the lower Coastal Plain is not shown, but we will cover Coastal Plain rocks in detail in this book. Here is a list of Georgia's physiographic provinces from northwest to southeast:

Cumberland Plateau Province (red tint on map)

Valley and Ridge Province (blue tint on map)

Blue Ridge Province (brown tint on map)

Piedmont Province (dark green tint on map)

Coastal Plain Province (yellow tint on map)

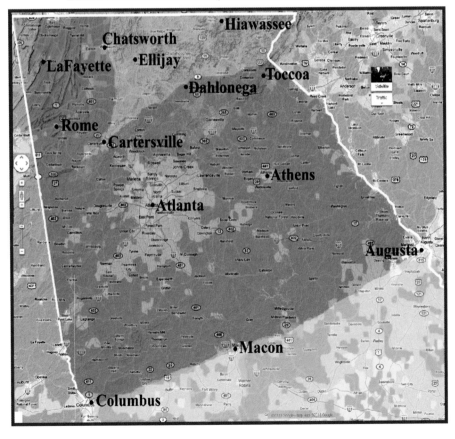

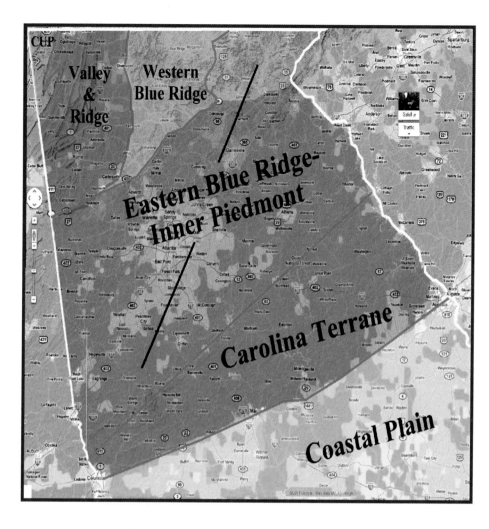

This map shows **red** lines that divide Georgia into six **geologic** provinces. You can see the physiographic and geologic provinces coincide to a great extent, but there are important differences.

How do the geologic provinces differ from the physiographic provinces? The **physiographic** provinces show land areas based on present-day land shape. The Cumberland Plateau is an elevated fairly flat plateau. The Valley and Ridge province is a great fertile valley crossed by low ridges. The Blue Ridge province is a region of rugged mountains. The Piedmont contains rolling foothills, and the Coastal Plain is a vast flat sandy plain. These land shapes are the result of relatively recent (in geologic time) gentle uplift, followed by weathering of the landscape.

On the other hand, the <u>geologic</u> provinces represent distinct land areas, each of which is the result of a specific geologic event. In other words, each geologic province has a unique geologic history.

Let's examine this idea more closely. The **Cumberland Plateau (CUP)** and the **Valley and Ridge (V&R)** provinces were once situated near the eastern edge of ancient North America (often called Laurentia by geologists). They became basins, receiving sediments eroded from three successive ancient mountain chains that we will discuss in detail. Subsequent folding, uplift and erosion sculpted the Cumberland Plateau into Sand, Lookout and Pigeon mountains, leaving the Valley and Ridge province a broad fertile valley, crossed by several low ridges.

The **Western Blue Ridge (WBR)** was also once a basin created when the eastern edge of the continent rifted apart between 750 and 550 million years ago (MYA). The basin filled with sediments, and then later it was folded and shoved up on top of the Valley and Ridge province to become Georgia's Cohutta Mountains and the Murphy Marble belt.

The **Eastern Blue Ridge/Inner Piedmont (EBRIP)** began existence as a chain of volcanic islands that crashed into Laurentia around 500 MYA.

The **Carolina Terrane (CT)** was yet another volcanic island chain that crashed into the mainland around 400 MYA.

The **Coastal Plain (CP)** consists of sediment accumulated since the most recent rifting of the continent around 150 MYA.

The following chapters tell in detail the story of how Georgia's geologic provinces were formed, giving rise to the rocks we find here today. Here is the greatly abbreviated version:

- **Supercontinent Rodinia rifts apart, 750-550 MYA**

- **EBRIP crashes into Laurentia, 500 MYA**

- **Carolina Terrane crashes into Laurentia, 400 MYA**

- **Africa crashes into Laurentia completing the supercontinent Pangea, 300 MYA**

- **Pangea rifts apart, 150 MYA**

CHAPTER 2. PLATE TECTONICS,
A MOVING EXPERIENCE

Today we know **plate tectonics** drives virtually all of earth's geologic processes. This chapter summarizes those processes by presenting a brief description of plate tectonics.

Our earth is composed of giant crustal plates moving on a planet-size cauldron of hot plastic rock called the mantle located about 25 miles below earth's surface. The crust (where we live) is just the cooled top of the mantle. It is similar to the thin crust of ice on a winter pond.

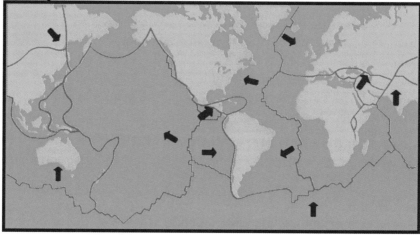

Earth's crustal plates outlined in red

Two kinds of earth crust exist; **ocean crust** and **land crust**. Ocean crust is dark in color and very heavy, so it sinks down, forming the ocean bottom. Land crust is light in color and light in weight so it rides above ocean crust, forming land. Earth's giant crustal plates are usually made of a combination of ocean and land crust.

There are really just two plate tectonic events that can happen to earth's giant crustal plates, **rifting** and **convergence**. These extremely important events together cause virtually all of the movement of the crustal plates, resulting in the creation of oceans and mountains. Both rifting and convergence are caused by **mantle hot spots**. Here's how.

As Earth's core slowly cools, mantle movements called convection currents ebb and flow. This pushes hot expanding <u>molten</u> mantle rock (**magma**) up under ocean crust, forming a hot spot hundreds of miles long. When a mantle hot spot develops, it makes a great bulge in the overlying ocean crust, creating a mid-ocean ridge. As pressure builds, hot magma burns through the ocean crust, spewing out on to the ocean bottom. Once magma breaks through the crust it is called **lava**. Lava continues to spew out in underwater volcanoes, actually creating new ocean crust that shoves existing crust out horizontally on each side of the volcano. As a result, the ocean grows wider.

A hot spot can also develop under the land crust of a continent. When this happens, the continent breaks apart and a new ocean develops between the two continent pieces. Whether the hot spot occurs beneath ocean crust or land crust, the resulting event is called **rifting** because there is a rift or split in the crust. **Rifting events create and expand oceans**.

For every action, there is an equal and opposite reaction. If rifting occurs on one side of a plate, creating new ocean; then, the extra ocean crust thus created must go somewhere, since earth's surface is just so big. It is like trying to fit the skin of an apple around a grape! On the opposite side of the plate from the rift, the excess ocean crust actually breaks in two, and one huge section of crust is shoved beneath the other. This is called **subduction**.

As more and more crust is subducted, it dives deep into the mantle where it melts into magma. All this extra magma creates great heat and pressure. Once again something has to give! Light weight magma that forms land crust separates from the magma blob of the subducted plate, and rises up through the overlying ocean crust to form a volcanic island. It is really an island chain since, just like hot spots, subduction zones form along lines many miles long.

The Aleutian Island chain off the coast of Alaska is a present-day example of a volcanic island chain formed by ocean plate subduction.

After a volcanic island chain is created, subduction can continue, shrinking the ocean basin between the island chain and a nearby continent until the island chain crashes into and becomes attached to the continent. This event is called **convergence** because two land masses converge on and crash into each other. Huge mountain chains are created in the process. **Convergence creates and expands continents**. Indeed, in the past, convergence has resulted in the formation of several gigantic continents called supercontinents. These supercontinents lasted for tens of millions of years, only to be eventually rifted apart again, forming several smaller continents separated by oceans.

The rifting and convergence events we've been describing actually define the boundaries of earth's crustal plates. The plate edges are huge linear zones of either rifting or convergence. Go back to the world map on page 6. The blue arrows point toward subduction zones where volcanic island chains are forming. The blue arrows point away from rifting zones where oceans are expanding.

The rocks of Georgia were formed by a series of rifting and convergence events occurring during the last billion years. These events were enormous in scope and size, affecting not only Georgia, but the entire eastern coast of North America. as well as the continents of Europe, South America, Asia and Africa.

On the next page you will find a Tectonics History Table listing tectonic events along America's east coast during the last 1 billion years. The next chapter tells this story in pictures.

Incidentally, as you read further, please note the dates for events described in this book are approximate. They have been "rounded" for simplicity and to reflect the uncertainty in projecting a timeline of such magnitude. Few geologists agree completely about them, and they will undoubtedly change as we learn more.

TIME	EVENT	LAND MASS
1 Billion Years Ago	Supercontinent Rodinia exists	Rodinia, with Laurentia part of its eastern edge
750 MYA (750-550)	Rodinia **rifts** apart in 2 phases over 200 million years	Ancient Africa and proto-**EBRIP** are rifted apart from Laurentia, **WBR** sediments deposited
600 MYA	Ocean crust subducted beneath ocean crust.	**Carolina Terrane** volcanic island is created near Ancient Africa
510 MYA	Ocean crust subducted beneath underwater island	**EBRIP created,** volcanic island (IP) and adjacent volcanic sediments
500 MYA	Taconic orogeny, EBRIP crashes into Laurentia	EBRIP now attached at the Hayesville Fault EBRIP erodes, depositing sediments on the future **CUP** and **V&R** provinces
400 MYA	Acadian orogeny, Carolina Terrane crashes into Laurentia plus EBRIP	**Carolina Terrane** now attached at the Middleton-Lowndesville Fault Few sediments remain from erosion of the CT
300 MYA	Alleghany orogeny, Ancient Africa crashes into Laurentia plus EBRIP plus Carolina Terrane.	The supercontinent Pangea is formed. Tremendous thrusting, folding and faulting occurs, creating metamorphic rocks.
150 MYA	Pangea rifts apart,	The Atlantic ocean is formed with proto-North America on the west and Africa on the east. Seas cover the lower CT forming the **Coastal Plain**

CHAPTER 3. PLATE TECTONICS HISTORY OF GEORGIA

Welcome to the east coast of supercontinent Rodinia around 1 billion years ago. Rodinia is made of several continents including Laurentia (ancient North America) and ancient Africa (also called Gondwana). A light blue sky presides over a dark blue ocean. The tan texture represents land crust. The black texture represents ocean crust. The hot plastic mantle is orange.

A mantle hot spot has developed beneath Rodinia. The hot spot is miles long and oriented on a northeast-southwest axis (based on current world arrangement). The land crust above the hot spot is bulging up, thinning and cracking. The hot spot will burn through Rodinia, breaking off several hunks of it, and creating a new ocean in the process. Laurentia and Africa will soon be separated, with the Iapetus (ancient Atlantic) ocean in between.

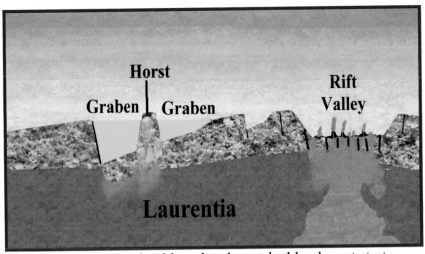

As rifting continues, the thinned and stretched land crust starts cracking. Hunks of land crust near the hot spot break loose and sink down to form basins called <u>grabens</u>. Sediments (yellow) from adjacent high spots called <u>horsts</u> have partially filled two grabens. Many such grabens are created.

A rift valley has formed on the right where volcanoes spew lava on the valley floor. The new ocean will develop here as rifting continues and the sea pours in. The entire rifting process will take about 200 million years!

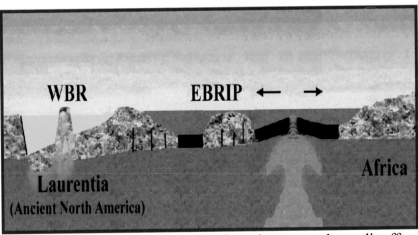

Rifting has continued, and a hunk of continent crust has split off and drifted a short distance away as an underwater island off Laurentia's new coast. The hot spot is now to the right (east) of the island. It is creating a new ocean between Laurentia and Africa.

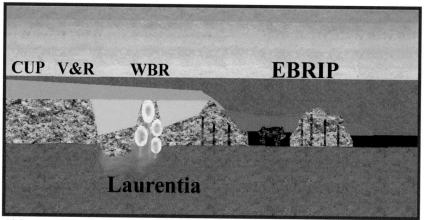

This closeup shows that rifting is now complete. The new ocean to the east is complete. Sediments (yellow) have completely filled two grabens, and limestone and associated sediments (pink) have covered everything. The Cambrian explosion of life will occur around 530 MYA, and its fossils will form in the pink limestone. The continent edge has sunk below the sea.

The underwater island plus sediments between it and Laurentia will become EBRIP (Eastern Blue Ridge-Inner Piedmont) geologic province. The two WBR grabens will become the Western Blue Ridge province. Land west of the WBR will become Georgia's Valley and Ridge province and its Cumberland Plateau province.

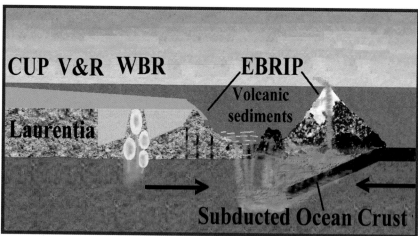

The time is now about 510 MYA. Somewhere far from this picture rifting is opening up a new ocean. Pressure from this expansion has caused a hunk of ocean crust to be subducted beneath EBRIP creating a volcanic island and associated volcanic sediments.

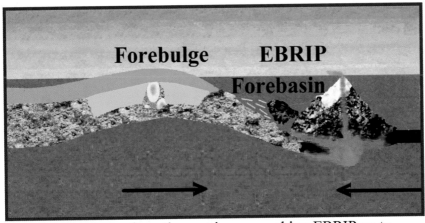

Continued pressure from the east is now pushing EBRIP on to
Laurentia. The pressure from EBRIP has created a deep basin
(forebasin) between it and Laurentia. It has also caused land west
of the basin to bulge up above ocean level (forebulge). This
forebulge in combination with other events will have catastrophic
effects on sea life around the world, and is described later.

EBRIP has crashed into Laurentia. This occurred about 500 MYA
and is called the Taconic orogeny. Orogeny means mountain
building and this convergence event is creating mountains. The
picture shows phase I of the Taconic orogeny. EBRIP has been
pushed on top of the first WBR graben and its sediments.

When EBRIP is shoved up on Laurentia, it pushes Laurentian crust
down into the mantle where it melts and rises as granite plutons
emplaced in EBRIP . When Africa crashes the granite plutons will
be metamorphosed into granite gneiss and eventually be exposed
as EBRIP's many granite gneiss plutons

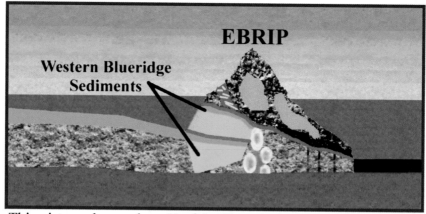

This picture shows phase II of the Taconic orogeny. Here, the first WBR graben and EBRIP have been pushed on top of the second WBR graben, creating a huge mountain range called the Taconic Mountains.

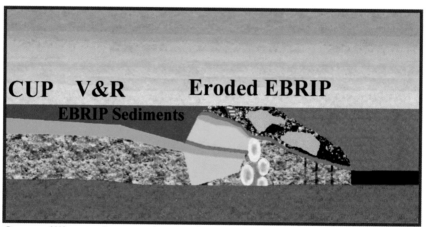

Over millions of years, the Taconic Mountains (EBRIP) have eroded down to sea level. All of the erosion sediments (brown) are deposited in a massive basin that will eventually become the Valley & Ridge province and the Cumberland Plateau province.

Nearly all the rocks of the WBR and EBRIP are now in place, but they don't look like today's rocks because they have experienced only minor amounts of the **intense deformation that will turn them into metamorphic rocks** (P.17). This will happen when Carolina Terrane and then Africa crash into the continent. Huge sheets of rock will be thrust on top of each other and pushed into folds. The folds will then be broken and one fold section shoved on top of another. Metamorphic rocks will emerge.

The Carolina Terrane (CT) was also a volcanic island chain like EBRIP. It formed when one ocean plate was subducted beneath another. It converged on Laurentia and crashed into it around 400 MYA in the Acadian orogeny. The oblique angle of the Carolina Terrane's collision with Laurentia did not produce high mountains or an extensive sediment basin like those of the Taconic orogeny, so few of its erosion sediments remain today.

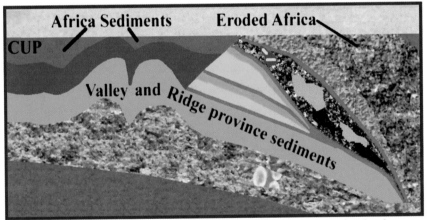

Here is the third and most massive convergence event in the last billion years, the Alleghany orogeny. The continent of Africa (also called Gondwana) crashed into Laurentia about 300 MYA. It slid on top of the Carolina Terrane, and shoved the Western Blue Ridge plus EBRIP, plus CT miles on top of the Valley and Ridge province (pink). Then it folded and finally broke those rocks, turning most of them into metamorphic rocks. Eventually, Africa eroded to sea level. Most of its sediments (gray) went into the basin that will become the Cumberland Plateau (CUP).

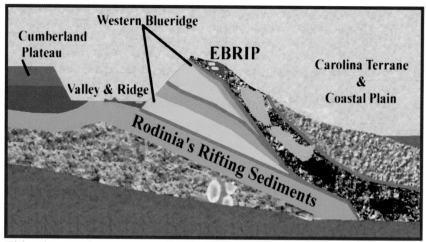

This picture shows generally the Georgia of today (vertical scale exaggerated), **not including all the folding and faulting**. What happened to Africa? The supercontinent formed when Africa crashed into Laurentia is called Pangea. About 150 MYA Pangea rifted apart. This rifting event, shoved the Atlantic ocean in between Laurentia and Africa, pushing Africa to its approximate present location. Only Africa's sediments (gray) remain.

Today's Cumberland Plateau is made of the sediments of Africa (gray) on top of EBRIP sediments (brown) which lie on top of limestone and associated sediments from Rodinia's rifting (pink).

The Valley and Ridge province is composed mostly of EBRIP sediments (brown) on top of sediments from the rifting of Rodinia (pink).

Today's Western Blue Ridge Mountains are made of sediments (yellow) deposited in the two grabens when Rodinia rifted, and later shoved miles up on top of sediments from Rodinia's rifting (pink).

Today's EBRIP began as a rifted hunk of Laurentia/Rodinia along with adjacent sediments off Laurentia's coast. Subducted ocean crust turned the rifted hunk into a volcanic island and caused the the sediments to be altered by volcanic activity. This whole package became EBRIP.

The Carolina Terrane became the lower Piedmont province and the underlying foundation of the Coastal Plain province.

CHAPTER 4. ROCK PRIMER

Before we go further into Georgia's rocks, a basic look at rocks in general is in order. Most folks know there are three main types or classes of rocks, **igneous**, **sedimentary** and **metamorphic**. Let's review each of these major rock classes.

Igneous rocks (p. 77) come directly from earth's molten mantle. Molten rock from the mantle is called magma. Magma rises from mantle hot spots, or it rises from subducted ocean or land crust. If the magma erupts on to the surface either above or below ocean surface as a volcano, it is called lava. The dark basalt lava of Hawaii is a familiar example.

Sometimes lava is ejected from a volcano with such force it shatters into pieces of all sizes that settle on the volcano's flanks as pyroclastic rock. Pyro means fire in Greek, and clastic means broken. Both EBRIP and CT produced abundant pyroclastic rock during their volcanic island phases. The metamorphosed versions of these rocks are all around us today.

If magma spews into deep cracks and pockets but never breaks the surface, it cools to form an intrusive rock body called a pluton, or if very large, a batholith. Stone Mountain is a great example of a granite pluton. Originally Stone Mountain formed deep underground, but erosion over many, many years has exposed it.

Sedimentary rocks (p. 83) form when igneous or metamorphic rocks are broken up and transported by wind, water and ice to form layers that harden into rock. This weathering process usually sorts the layered material by size, so you get sandstone (sand-size particles) from former river margins, beaches and deltas. Smaller silt and mud-size particles wash into shallow water where they form mud flats that harden into shale. Coral reefs form in shallow warm seas. Calcium from the reef body and associated marine life settles on the bottom to harden into limestone and dolomite (if magnesium is present in the limestone).

Metamorphic rocks (p. 98) form when igneous and sedimentary rocks are exposed to tremendous heat and pressure over long periods of time. This results from folding, faulting and burial of rock layers during convergence and crashing events. A metamorphic rock can be moderately metamorphosed or highly metamorphosed, depending on how much metamorphic deformation it has suffered.

During metamorphism, light colored minerals like quartz and feldspar often migrate into bands and swirls in the rock. Continental movements involving sliding and folding can create a metamorphic rock that is layered (foliated) and looks like a sedimentary rock. Hot ocean water from volcanic vents can pulse through a porous igneous rock, depositing new minerals in a metamorphic process called hydrothermal alteration.

What Rocks Are Made Of

An element is a pure substance. It can not be divided into different parts, either physically or chemically. Silicon and oxygen are examples of elements. Put two or more elements together and you get a mineral. A mineral can be divided into parts chemically, but not physically. Put two or more minerals together and you get a rock.

Cheeseburgers and rocks have a lot in common! A cheeseburger can be broken down physically into its parts (bun, meat, cheese, pickle). Similarly, a rock like granite can be physically broken down into its minerals (quartz, feldspar, muscovite, biotite). Chop a piece of cheese into a thousand pieces, and each one is still cheese. Chop a piece of quartz into a thousand pieces, and each one is still quartz. Is this making you hungry?

Minerals and rocks can be **felsic, intermediate, mafic** or **ultramafic** in reaction, depending on the elements they contain. Let's focus more closely on these extremely important terms.

Felsic refers to minerals that contain light colored, light weight elements with relatively low melting temperatures found abundantly on land. Silicon, oxygen, aluminum and potassium are elements that combine to form felsic minerals like quartz, potassium feldspar and muscovite mica.

Mafic refers to minerals that contain abundant dark colored, heavy elements with relatively high melting temperatures found abundantly in the upper half of ocean crust. Iron, magnesium and calcium are elements that combine to form mafic minerals like hornblende and calcium plagioclase. Mafic minerals do contain some felsic elements like silicon, but not in large amounts.

Intermediate refers to minerals containing a mix of light and dark elements. The minerals formed are intermediate between felsic and mafic in their makeup. Biotite and sodium plagioclase are

18

examples of intermediate minerals.

Ultramafic refers to minerals that contain only the most mafic elements like magnesium and iron with virtually no felsic elements. Ultramafic minerals like pyroxene and olivine are found abundantly in lower ocean crust but rarely on land.

Minerals in Georgia Rocks

MINERAL NAME	ELEMENTS	FELSIC OR MAFIC, etc.	BRIEF DESCRIPTION
QUARTZ	Silicon, Oxygen	Felsic low melting temp.	* Quartz in rocks nearly always looks **dark gray or clear,** rarely white
POTASSIUM FELDSPAR	Silicon, Oxygen, Aluminum, Potassium	Felsic low melting temp.	* Potassium feldspar in rocks can look off white or pinkish, but usually it looks **white**
MUSCOVITE MICA	Same as K-spar, plus water as hydroxide	Felsic low melting temp.	* Muscovite in rocks looks like **small silvery specks or flakes.**
BIOTITE MICA	Same as muscovite, plus small amounts of Iron (Fe) and Magnesium (Mg)	Intermediate intermediate melting temp.	* Almost always found as **small black specks** in rocks instead of separate large crystals. * **A nickel (coin) can easily scratch biotite, making a gray dust (p. 115).**

MINERAL NAME	ELEMENTS	FELSIC OR MAFIC, etc.	BRIEF DESCRIPTION
PLAGIOCLASE FELDSPAR	Silicon, Oxygen, Aluminum, and some ratio of **Sodium** to **Calcium**	Intermediate to Mafic intermediate to high melting temp.	* Plagioclse feldspar looks **white in most rocks,** but in amphibolite and diabase rocks it can look clear
HORNBLENDE also called **AMPHIBOLE**	Lots of iron and magnedium, plus some Sodium and Calcium	Mafic high melting temp.	* Usually found as small black, needle-shaped or rod-shaped crystals in rocks. *Easily confused with biotite, but a nickel **cannot** scratch it (p. 115)
PYROXENE	A Little Silicon, Oxygen, and Aluminum. Lots of Calcium, Iron and Magnesium	Mafic to Ultramafic high melting temp.	* Black, greenish, gray, maroon or bluish. * Found rarely on land in suture faults and isolated patches called klippen
SERPENTINE	Olivine plus water(as OH)	Ultramafic very high melting temp.	* Greenish gray * Found rarely on land in suture faults

MINERAL NAME	ELEMENTS	FELSIC OR MAFIC, etc.	BRIEF DESCRIPTION
TALC	Serpentine plus carbon dioxide and water	Ultramafic very high melting temp.	* White or greenish * Fingernail can scratch it. * Found in suture faults & klippen
OLIVINE	90 % Magnesium	Ultramafic very high melting temp.	* Greenish gray * Rare and unstable on land.

Types of Feldspar Table (a confusing mineral!)

FELDSPAR NAME	ELEMENTS in addition to silicon and aluminum	FELSIC OR MAFIC, etc.	COMMON NAME
Potassium Feldspar	Potassium	F	Microcline (if it cooled slowly) Orthoclase (if it cooled quickly)
Sodium Feldspar	Sodium	I	Plagioclase (Albite)
Sodium-Calcium Feldspar	more Sodium than Calcium	M	Plagioclase (Oligoclase)
Sodium-Calcium Feldspar	Equal amounts of Sodium and Calcium	M	Plagioclase (Andesine)
Calcium-Sodium Feldspar	more Calcium than Sodium	M	Plagioclase (Labradorite)
Calcium Feldspar	Calcium	M	Plagioclase (Anorthite)

Here is a wacky way to remember the plagioclase feldspars in order of their sodium to calcium ratio:

I'll bite Ollie's and Andy's Labrador and or fight!

Many of Georgia's rocks contain one or more of the following minerals. If you can recognize these few minerals you are well on your way to identifying Georgia rocks.

Quartz, looks glassy and dark gray or clear in rocks

Quartz crystal, forms in rock pockets and big seams

Potassium Feldspar. Large specimens have a flat flashy face, but this is rarely seen in rocks. Looks white in rocks

Plagioclase Feldspar also has a flashy face. Looks white in rocks

Muscovite Mica, silvery specks in rocks

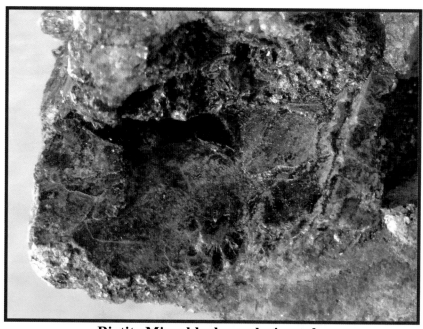

Biotite Mica, black specks in rocks

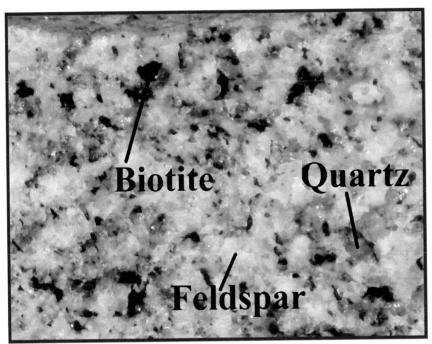

Minerals in Elberton Granite

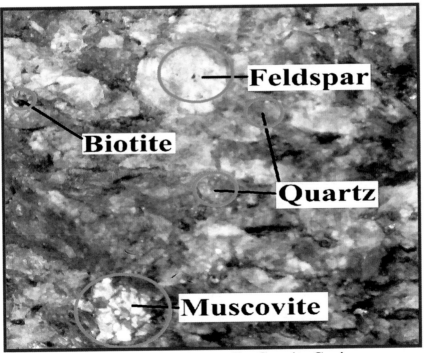

Minerals in Oglethorpe Co. Granite Gneiss

The following tables summarize igneous, sedimentary and metamorphic rock classification.

IGNEOUS ROCKS TABLE

EXTRUSIVE IGNEOUS ROCK (LAVA)	INTRUSIVE IGNEOUS COUNTERPART (PLUTONS)	FELSIC, INTERMEDIATE, MAFIC OR ULTRAMAFIC IN COMPOSITION
Rhyolite	Granite (77)	Very Felsic
Dacite	Granodiorite	Felsic
Andesite	Diortie	Intermediate
Basalt	Gabbro or Diabase (80)	Mafic
--------	Peridotite (122)	Ultramafic

If the above igneous rocks table confuses you, just remember, magma (plutons) and lava are made of the same stuff. Lava is just magma that comes out on the surface. You could just call it all "**magma/lava**." Based on the table above we can say there are generally four kinds of magma/lava:

* light colored magma/lava

* intermediate magma/lava

* dark colored magma/lava

* ultramafic ocean crust magma/lava

Granite, like that of Stone Mountain and diabase (a dark magma/lava emplaced during continental rifting), are examples of Georgia **igneous** rocks.

SEDIMENTARY ROCKS TABLE

SEDIMENTARY ROCK TYPE	SOURCE MATERIAL	PROVINCE
Conglomerate (83)	Layered Sand and pebbles	WBR, V&R, CUP
Turbidite Sandstone(83)	Mixed up Sand and pebbles	WBR
Sandstone (85)	Sand-size quartz particles	WBR, V&R, CUP, CP
Graywacke (88)	Mixed-size quartz and K-spar particles	WBR
Siltstone (103)	Silt-size quartz and feldspar particles	WBR
Shale (89)	Mud-size quartz and k-spar particles	CT, WBR, V&R, CUP, CP
Chert (91)	Microscopic quartz particles in beds and nodules	EBRIP , CT, V&R, CUP
Kaolin (93)	Feldspar washed into Fall Line swamps	CP, Fall Line
Limestone (94)	Calcium from coral reefs and shell creatures	V&R, CUP, CP
Dolomite (94)	Limestone with magnesium added	V&R
Coal (97)	Peat from swamps	CUP

Sedimentary rocks are widely used by people. Sandstone from the Cumberland Plateau can be found in dry-stack stone walls or house fronts. Shale and its metamorphic equivalent slate are used for paving stones, floors and specialty house roofs. American Indians made blades and arrow points from chert and its close relative flint. Limestone products make our lawns and gardens grow better, and constitute a main ingredient in the making of concrete. Coal powers factories.

METAMORPHIC ROCKS TABLE

METAMORPHIC ROCK TYPE	ORIGINAL ROCK	ORIGINAL ROCK CATEGORY
Quartz Veins and crystals (98)	High-silicon host rocks	All three classes
Quartzite (100)	Sandstone	Sedimentary, felsic
Granofels (101)	Amphibolite	Metamorphic
Metachert (102)	Chert	Sedimentary, felsic
Slate (103)	Shale	Sedimentary, felsic
Phyllite (103)	Slate	Metamorphic, felsic
Quartz Muscovite Schist (105)	Mixed sand to mud-size material	Sedimentary, felsic
Muscovite Schist (108)	Mud-size felsic pyroclastic material	Sedimentary felsic
Metadacite (110)	Dacite lava	Igneous, felsic
Granite Gneiss (112)	Granite pluton	Igneous, felsic
Biotite-Hornblende Gneiss (115)	Andesite lava or Diorite pluton	Igneous, intermediate
Amphibolite and Metagabbro (118)	Basalt lava, Gabbro pluton or diabase dike	Igneous, mafic
Weathered Mafic Rock (121)	Mafic pyroclastic rock?	Igneous
Serpentinite, Soapstone (122)	Peridotite	Igneous
Marble (125)	Limestone	Sedimentary

The word **gneiss** means a metamorphic rock high in quartz and feldspar that often shows metamorphic banding. The word **schist** means a metamorphic rock in which soft minerals like muscovite, biotite, chlorite and talc are flattened and aligned parallel.

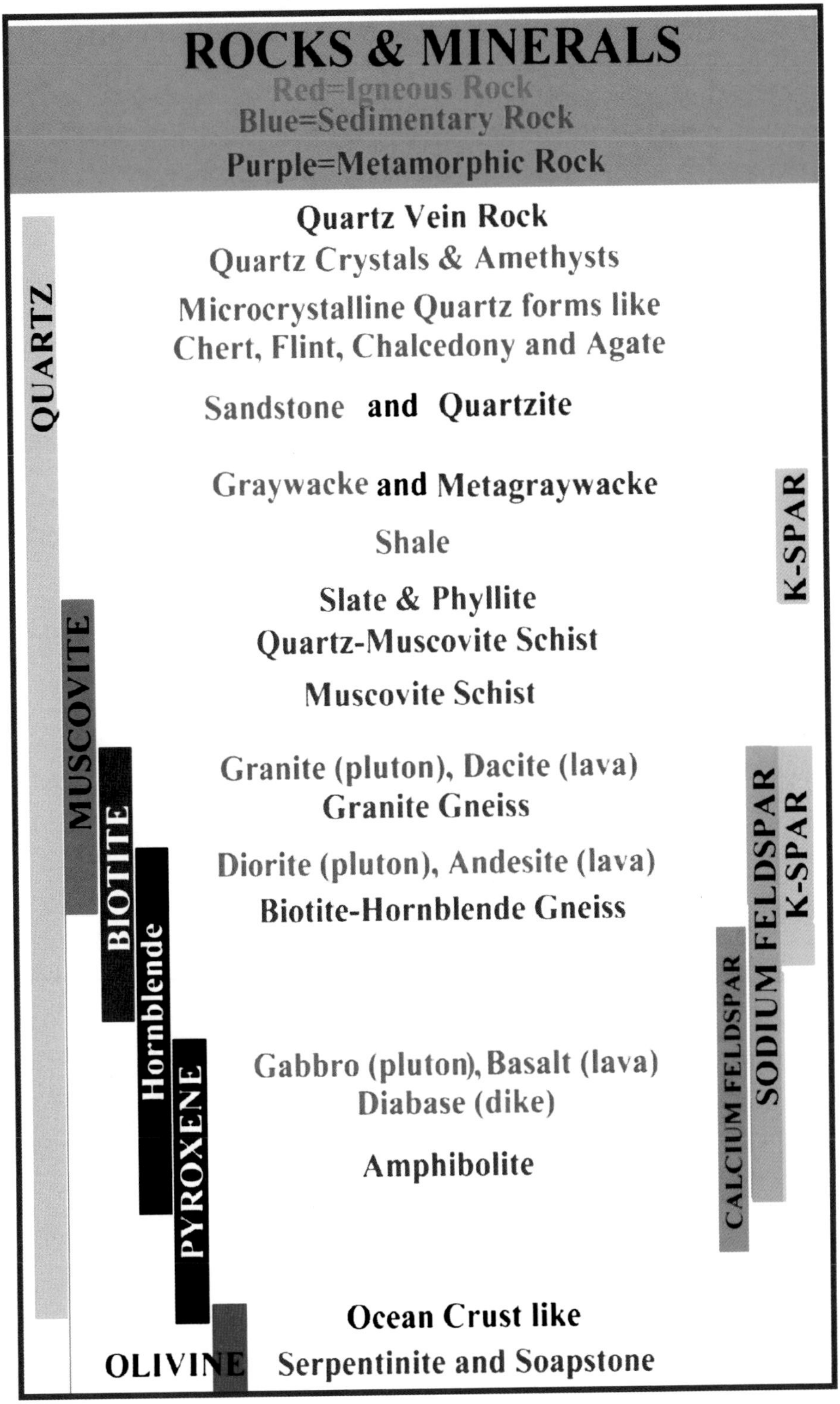

ROCKS & MINERALS

Red=Igneous Rock
Blue=Sedimentary Rock
Purple=Metamorphic Rock

QUARTZ

Quartz Vein Rock

Quartz Crystals & Amethysts

Microcrystalline Quartz forms like
Chert, Flint, Chalcedony and Agate

Sandstone **and** Quartzite

Graywacke **and** Metagraywacke

Shale

Slate & Phyllite
Quartz-Muscovite Schist

Muscovite Schist

K-SPAR

MUSCOVITE

BIOTITE

Hornblende

PYROXENE

Granite (pluton), Dacite (lava)
Granite Gneiss

Diorite (pluton), Andesite (lava)
Biotite-Hornblende Gneiss

CALCIUM FELDSPAR

SODIUM FELDSPAR

K-SPAR

Gabbro (pluton), Basalt (lava)
Diabase (dike)

Amphibolite

Ocean Crust like

OLIVINE **Serpentinite and Soapstone**

CHAPTER 5. WESTERN BLUE RIDGE (WBR)

We begin our detailed descriptions of Georgia's geologic provinces with the **Western Blue Ridge** because its rocks were the earliest formed. By about one billion years ago, the supercontinent Rodinia had been assembled by a series of convergence (mountain building) events and was eventually eroded down to sea level.

This eroded platform of land crust was made of <u>granite gneiss</u> (p. 112), the light colored, light weight felsic rock that composes continental crust. Geologists call this granite gneiss, **Grenville basement rock** because it forms the roots of the ancient eroded Grenville Mountains that helped create Rodinia. It still underlies all of Georgia's rocks as augen granite gneiss (p. 114).

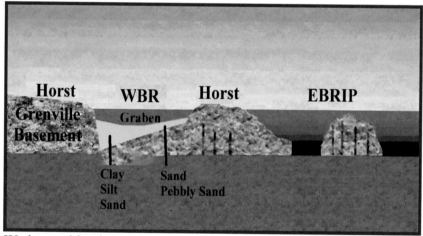

We learned in chapter 3 that Rodinia began to rift apart around 750 MYA. The hot spot that did the rifting caused the continental crust to be thinned and stretched. Huge blocks of the crust broke loose and slumped down creating basins called grabens surrounded by high spots called horsts. The WBR graben, oriented longways from northeast to southwest, was several hundred miles long, stretching from about Asheville, North Carolina through Georgia.

Now let's look more closely at the sediments (yellow) that flowed into the WBR graben. Geologists call them the **Snowbird Group** of sediments. Most of these sediments came from the horst on the right in the picture. Think of a river flowing from the horst and into the water-filled graben. Coarse sediments like pebbles and sand would be dropped by the river as it first entered the graben, forming river mouth and river delta deposits. Finer particles like clay would be carried farther out (northwest) into the graben and would pile up thicker than the sand. This is exactly how these sediments formed. The mixed pebbles and sand hardened into a rock called conglomerate (p. 83). The sand hardened into sandstone (p. 85). The clay became shale (p. 89).

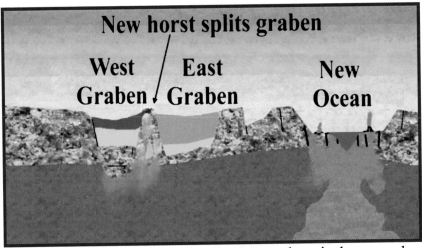

After the Snowbird sediments (yellow) were deposited, a second more rapid phase of rifting occurred in which a new horst rose up and split the WBR graben in two. Let's call the graben on the left West Graben and the graben on the right East Graben. As the new horst rapidly rose, East Graben rapidly sank, becoming much deeper with steep underwater sides. West Graben sank more slowly.

The picture above shows that sediments from the new horst have filled both basins. The West Graben green sediments are called the **Walden Creek Group**, and the East Graben orange sediments are called the **Great Smoky Group**. Both groups were deposited on top of the yellow Snowbird Group.

Most of the orange sediments, made of pebbles, sand, silt and clay were deposited very rapidly, pouring into East Graben, where they tumbled down its underwater sides, getting all mixed up. Deposits of this kind are called turbidites. Small pieces of a **blue-colored quartz** (p. 84) from ancient Grenville basement rock pulled up from below into the middle horst got mixed in with some of these sediments, providing an identification clue for these rocks.

This closeup shows the final stage of the rifting of Rodinia, (about 550 MYA) where the continent edge has subsided below the sea, and limestone and associated sediments (pink) have covered everything. The table below shows the graben sediments and the limestone and associated sediments that covered them in chronological order with the oldest on the bottom. The graben sediments will eventually become the rocks of the Western Blue Ridge province.

West Graben Sediments	East Graben Sediments
Knox Limestone Conasauga Group Rome formation Chilhowee Group	Murpy Marble Brasstown Formation Tusquiti Quartzite Nantahayla Formation
Walden Creek Group	Great Smoky Group
Snowbird	Snowbird

At the edge of the sea you have sandy beaches and river deltas (sandstone). In shallow waters just off shore, mud flats form (shale) and farther out, coral reefs (limestone) form. When the sea rises, as happened when Rodinia rifted, first mud (shale) covers the sand and then coral reefs (limestone, p. 94) cover the mud. If you were to look at, say a road cut that exposed sediments laid down when the sea was advancing, you would see a layer of sandstone on the bottom with a layer of shale next and finally a layer of limestone on top. If the sea rose in fits and starts instead of steadily, you would have alternating layers of the three rocks covered by a <u>final layer of limestone</u> marking permanent submergence. The rocks in the purple row of the table (Knox, etc. and Murphy, etc.) show exactly this sequence. The Knox Limestone and Murphy limestone layers mark the time when Rodinia's edge finally sank permanently under the sea.

As you already know, about 500 MYA the Taconic orogeny occurred, in which the EBRIP volcanic island and sediments crashed into Laurentia. First, EBRIP was pushed on top of East Graben, then both East Graben and EBRIP were pushed on top of West Graben forming the Taconic Mountains, a mountain chain as high as today's Rockies. Over time, the mountains eroded completely down to sea level. Then the Carolina Terrane island chain crashed into Laurentia and was, in its turn, eroded to sea level.

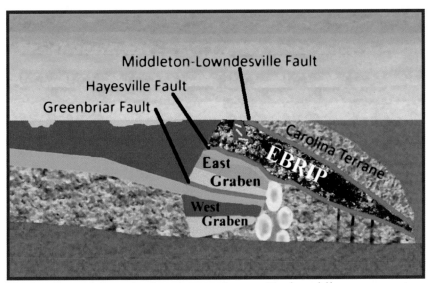

Notice the red lines in the picture above. Each red line represents a <u>thrust fault</u> where one hunk of land was thrust on top of another. The **Greenbrier Fault** marks the place where East Graben was thrust on top of West Graben during phase II of the Taconic orogeny. The **Hayesville Fault** marks the place where EBRIP was thrust on top of East Graben during phase I of the Taconic orogeny. You can trace its length on the Georgia map on page 4. It runs generally through Hiawassee, Dahlonega, Dawsonville and Canton, separating the EBRIP from the WBR. Some of the subducted ocean crust between EBRIP and Laurentia was caught up in the Hayesville Fault, resulting in some unique rock types like serpentinite and soapstone (p. 122).

The **Middleton-Lowndesville Fault** may also be traced on the Georgia map, representing the place where the Carolina Terrane volcanic island chain was thrust on top of EBRIP. It runs from near Richard Russell State Park, southwest across the state.

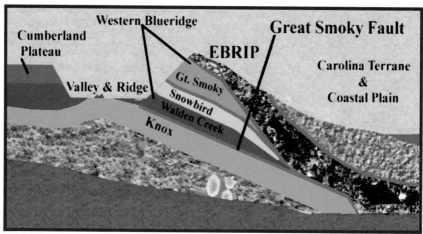

This final picture shows a side view of Western Blue Ridge rocks today. Thrusting, folding, faulting and erosion of WBR rocks resulted in the left-to-right arrangement of the Walden Creek, Snowbird and Great Smoky rocks depicted in the picture above; however, the picture (and all subsequent WBR pictures) does not show the many folds actually present in WBR rocks today.

There is one more red fault line located just below the WBR sediments. It is the **Great Smoky Fault**. This fault follows Georgia highway 411, and it marks the place where WBR plus EBRIP plus CT were thrust miles on top of the Valley and Ridge sediments (pink). Africa did this final job of thrusting, pushing this literal mountain of rock several miles across the top of the Knox Group of the V&R province. It also greatly folded, faulted and metamorphosed all the rocks east of the Great Smoky Fault.

After Africa rifted away for the final time, breaking up the supercontinent Pangea, gentle uplift raised the continent, and erosion carved streams into the land surface, creating outstanding scenic features in the WBR like Fort Mountain, the Cohutta Mountains, Rich Mountain, Coopers Creek and Amicalola Falls.

Let's now have a look at the United States Geologic Survey geology map of Georgia (http://mrdata.usgs.gov/sgmc/ga.html).

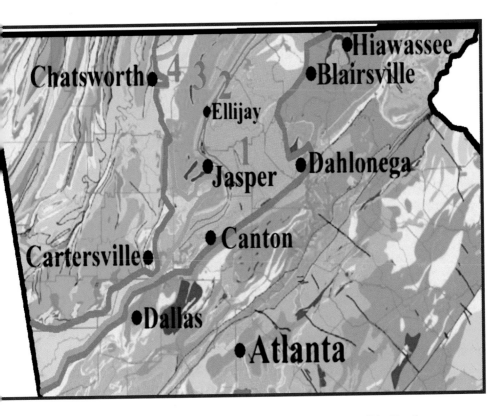

The red line on the right of this map marks the Hayesville Fault, running through Hiawassee, Dahlonega, Canton and Dallas. The red line on the left marks the Great Smoky Fault, through Chatsworth and Cartersville. The Western Blue Ridge province lies between the red lines.

Remember the Snowbird Group and the Walden Creek Group that were deposited in the West Graben (p. 31)? The orange line on this map follows the Greenbrier Fault that separates these rocks from those of the East Graben to the right. The Walden Creek rocks, are in the upper left of the map (north of Chatsworth), and the Snowbird Group rocks form a tail extending out to the west down below.

The Great Smoky Group rocks, deposited in the East Graben, make up the vast majority of Georgia's Western Blue Ridge rocks today. They are represented on the map by the four red numbers. The big red 1 (gray color zone) denotes the Great Smoky Group Dean Formation. The red 2 (lavender color) denotes the long slender

Murphy Marble Belt. The 3 (bright green) denotes the Great Smoky Group <u>Copperhill Formation</u>, and the red 4 (light purple) denotes the Great Smoky Group <u>Thunderhead Sandstone Formation</u>. The table on page 37 describes the rocks in each of these formations.

The image and table on page 32 show that Murphy rocks were once a <u>layer</u> of limestone and associated rocks at the continent edge. During the Alleghany orogeny, the Murphy layer was pushed into a huge downward fold called a syncline. The long narrow lines of today's Murphy Belt are the syncline edges.

WBR rocks experienced folding and faulting that turned them into metamorphic rocks. The continental pushing and shoving responsible for this was more severe in the southeastern part of the WBR (closer to smashing Africa!) and least severe in the northwest. That's why there is more <u>muscovite schist (p. 108)</u>, the highly metamorphosed version of clay sediments in the southeastern WBR and more slate (p. 103) and phyllite (p. 103), the less metamorphosed version of clay sediments, in the the northwest WBR.

As rifting sediments like those of the WBR accumulate, clay-size particles of quartz and feldspar wash down into basins to form mud flats and salt marshes. Metamorphic pressure first turns the mud into shale (a sedimentary rock). Further metamorphism turns shale into slate. Slate gets turned into phyllite, and finally, with continued metamorphism, phyllite becomes schist.

An interesting feature on the WBR geologic map is the purple color included in the Snowbird Group sediments just northeast of Cartersville between the Greenbrier and Great Smoky faults. The purple represents granite gneiss plutons. Unlike EBRIP's granite gneiss, these plutons are actually hunks of the ancient Grenville basement rock mentioned earlier (p. 30). Thrusting and erosion of overlying Snowbird rocks have exposed this 1 billion year old rock in what geologists call a geologic window. The two outcrops of this rock, shown in purple on the map are the Corbin and Salem Church granite gneisses. Severe metamorphism has mashed this granite's quartz and feldspar crystals into eye shapes geologists call augens, giving the rock the name, augen gneiss (p.114).

The table below shows WBR rock formations, the rock types in each formation with the main rock type underlined, and the page where each rock type is pictured and described.

ROCK FORMATION	ROCK TYPES	PAGE NUMBER
Murphy Group	<u>Marble</u> Muscovite Schist Quartzite Slate	125 `108 100 103
Great Smoky Group Dean Formation	<u>Muscovite Schist</u>	108
Great Smoky Group Copperhill Formation	<u>Metagraywacke</u> Muscovite Schist Amphibolite	88 108 118
Great Smoky Group Thunderhead Sandstone F.	<u>Turbidite Sandstone</u> Slate Quartzite	83 103 100
Walden Creek Group	<u>Phyllite</u> Quartzite Metagraywacke	103 100 88
Snowbird Group	<u>Quartzite</u> <u>Phyllite</u> Granite Gneiss	100 103 112

Notice the <u>Knox Limestone</u> is missing from this table and the WBR. It's because Africa shoved WBR rocks miles on top of the Knox limestone, which remains buried beneath WBR rocks today.

Follow Western Blue Ridge rocks northeast and you will find yourself in Great Smoky Mountains National Park along the Tennessee, North Carolina border. WBR rocks are well exposed and complete in the Park, and geologists were able to first understand their origin there.

Incidentally, geologists (Unrug & Unrug, 2000) recently thought they had discovered fossils in the Walden Creek Group rocks that lived at a much later date than that of the rifting of Rodinia. Subsequent research, however, revealed this to be untrue. We just wanted you to know in case you encounter it in conversation at your next pinata party!

CHAPTER 6. EASTERN BLUE RIDGE & INNER PIEDMONT (EBRIP)

The **Eastern Blue Ridge Inner Piedmont** (EBRIP) consists of the mountains of northeast Georgia and most of the Piedmont physiographic province. The Hayesville fault bounds EBRIP on the northwest, and the Middleton-Lowndesville fault bounds it on the southeast (map, p. 4).

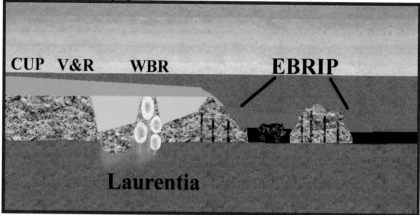

We already know from previous chapters that EBRIP began as a hunk of Laurentia that rifted off and drifted a little to the southeast. It also included the sediments between Laurentia and the rifted hunk. It looked like the picture above around 530 MYA.

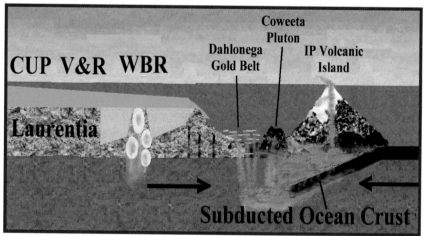

Around 510 MYA pressure from new ocean crust at a mid-ocean ridge expanding far to the east, caused ocean crust just east of EBRIP to break and descend below it. Three important geologic structures (also called terranes) were created by the subducting

38

ocean crust. First, the subducting ocean crust turned the rifted hunk into a **volcanic island** (really a chain of islands all along the Laurentian coast similar to today's Aluetian Island chain). Second, just west of the developing volcanic island, a <u>back-arc volcanic basin</u> was formed where hot magma circulating from the subducting ocean crust formed a large pluton (dark in the picture) that geologists include in the **Coweeta Group** of rocks. Finally, west of the Coweeta pluton, hot ocean water descended into magma chambers bringing up gold into the ocean floor right next to Laurentia, forming the **Dahlonega Gold Belt**.

These 3 terranes (Dahlonega Gold Belt + Coweeta Pluton + volcanic island) plus additional surrounding sediments make up today's EBRIP. Remember, EBRIP stands for Eastern Blue Ridge and Inner Piedmont.

This closeup of the island part of EBRIP shows in <u>simplified</u> form how the subducting ocean crust turned this part of EBRIP into a volcanic island. This very important process is called **fractional melting**, and here's how it happens.

When ocean crust is subducted, lots of ocean water goes down with it. Ocean water helps make fractional melting possible by lowering the melting temperatures of minerals it contacts.

The descending ocean crust and water heat up until the relatively low melting points of felsic minerals like quartz and potassium feldspar are reached. These light weight, easily melted minerals separate from the larger body of descending ocean crust and rise up, becoming emplaced as <u>felsic magma</u> (white texture) in the top of the EBRIP volcano.

As melting continues, the melting temperatures of intermediate minerals like biotite and sodium feldspar are reached, giving rise to an <u>intermediate magma</u> (gray speckled texture) that separates and rises up beneath the felsic magma.

Fractional melting can happen yet again, resulting in a <u>mafic magma</u> (dark texture) forming below the intermediate magma. The leftover ultramafic magma is so heavy with iron it sinks down and is incorporated back into the mantle. **For a refresher course on the terms felsic, intermediate, mafic and ultramafic, see page 18.**

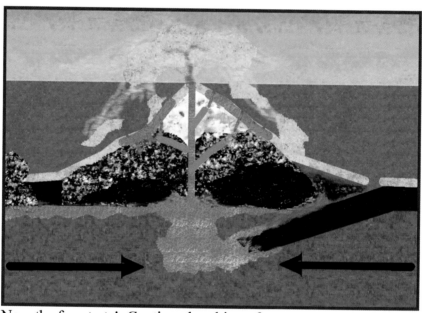

Now the fun starts! Continued melting of ocean crust creates great pressure from below. The felsic magma (white texture) at the top is very stiff, unable to flow easily like its more mafic counterpart due to strong silicon-oxygen bonds in abundant quartz. Instead, it explodes into millions of pieces of pyroclastic rock which flow down the volcano's underwater flanks. This pyroclastic rock consists of felsic material (quartz and feldspar) of all sizes from pebbles to dust that form a sedimentary rock called <u>felsic tuff</u>. (yellow texture) covering the volcano's flanks.

Water currents and gravity sort this material by size, with coarser rock and sand-size material on the volcano upper flanks and smaller silt and mud-size material farther out.

Super hot ocean water from volcanic vents flushes through some of the felsic tuff in a process called hydrothermal alteration, that leaches out sodium and calcium, and often adds interesting elements like iron, sulfur, copper, titanium and gold. The iron and sulfur (iron pyrite) impart a reddish tinge to portions of the tuff (reddish texture) that you can see in these rocks today.

The finest pyroclastic ash particles from the explosions settle far out, away from the volcano to become shale that is high in silicon, aluminum and potassium and even finer chert (microscopic quartz).

Some felsic magma may actually flow stifly from the volcano as dacite lava (light colored lava) following the pyroclastic explosion. It hardens into dacite rock (p. 110) containing large crystals (phenocrysts) of quartz and feldspar in a finer matrix of the same minerals.

Most of the intermediate and mafic magmas produced by fractional melting form plutons in the core of the volcanic island. The plutons are composed of diorite (intermediate) and gabbro (dark), respectively; however, in a few places on the volcano's flanks, these intermediate and mafic magmas break the surface as flows of andesite (intermediate) lava and basalt (dark) lava.

The explosive eruptions pictured in the image above happened over and over as the volcanic island developed. This resulted in a layer cake effect (not shown in the picture), with layers of felsic tuff alternating with layers of plutons and lava flows. Layer upon layer, the Taconic Mountains rose to great heights all along a chain of volcanic islands that stretched for hundreds of miles off the coast of Laurentia.

The Coweeta pluton was emplaced as diorite on the opposite side of the volcanic island from the subduction zone. This west side of the island is called a **back-arc** basin. Here, ocean water descended down into hot volcanic zones where it picked up gold and deposited it in the sands of the back-arc basin. Later, when EBRIP and Africa crash into Laurentia, water circulating through the sands will pick up the gold and silicon, redepositing them in large gold-laden quartz veins to create the fabulous Dahlonega gold region.

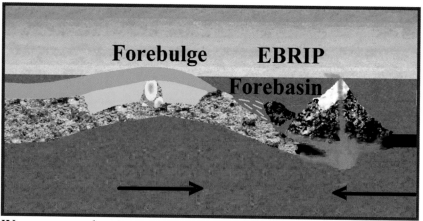

We are zoomed out now so we can see EBRIP and Laurentia. The Taconic orogeny is in full swing, and the 3 parts of EBRIP are colliding with Laurentia. Many geologist believe at this time the subduction zone changed, and ocean crust between Laurentia and EBRIP was subducted down and to the east (east-dipping subduction zone) under EBRIP. Others believe the west-dipping subduction zone remained. Either way, EBRIP's 3 parts were eventually pushed up on to Laurentia.

The Dahlonega Gold Belt (brown zone with yellow stripes) hit Laurentia first. Then the Coweeta Pluton was shoved up on top of the Gold Belt. Next came ocean sediments that had been deposited between the Coweeta pluton and the volcanic island, and finally, the volcanic island crashed into Laurentia.

A foreland basin (forebasin) has developed as EPRIP's parts push down on the edge of Laurentia. Land crust is flexible. EBRIP is pushing down on one end of the land crust, and the other end responds by rising up in a forebulge that lifts the edge of Laurentia above sea level. Behind the bulge (left), the continent bends back down into the sea.

As EBRIP moves closer to Laurentia, both the foreland basin and the forebulge move ahead of it until EBRIP finds its final resting place against Laurentia. At this point EBRIP's mountains cease to rise. Erosion begins and the eroded sediments fill the foreland basin until EBRIP is eroded to sea.

What about the forebulge? The bulge moves along ahead of EBRIP and the foreland basin until eventually it has raised up and then lowered the entire edge of Laurentia. This happened along thousands of miles of coast land as huge volcanic island chains

42

associated with EBRIP crashed into the supercontinent Pangea that was taking shape.

Virtually all life on earth at that time lived in the warm shallow seas on the edge of continents. The moving forebulge raised this cradle of life above sea level resulting in extinction of many sea creatures. Geologists call this die-off the Post Knox Unconformity because erosion of the raised up land left no sediments recording the time length involved.

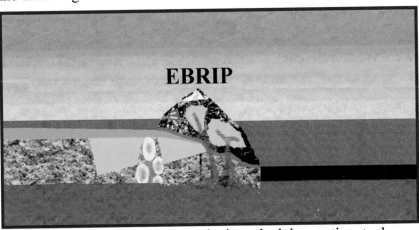

When EBRIP slid up on Laurentia, it pushed the continent edge down into the mantle. Continental crust melted and the resulting magma rose up into EBRIP as granite plutons. Later, when Africa crashed into the continent, the resulting heat and pressure turned the granite plutons into a metamorphic rock called granite gneiss, one of EBRIP's most abundant rock types today. Most of EBRIP's "granitic" rock outcrops are made of this granite gneiss instead of igneous granite.

Africa's crash gave rise to still more new granite plutons. Stone Mountain and Elberton granite are examples of granite plutons emplaced in EBRIP when Africa crashed. No crashing event has occurred since then, so these most recent plutons remain as granite instead of granite gneiss.

As previously mentioned, the crashing of Africa during the Alleghany orogeny (300 MYA) was massive in terms of the heat and pressure it applied to the rocks it was thrust upon. EBRIP's rocks were affected more than those of any other Georgia geologic province. Virtually all of EBRIP's rocks were folded, refolded, faulted (broken) and turned into metamorphic rocks.

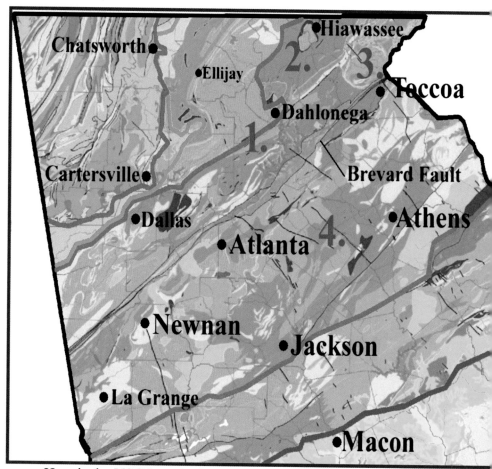

Here is the USGS geologic map of the EBRIP today. The red line running through Dahlonega marks the Hayesville Fault where EBRIP was attached (sutured) to Laurentia. The lower red line follows the Middleton-Lowndesville Fault. EBRIP lies between these two red fault lines.

The green numbers mark the 3 parts of EBRIP in addition to ocean sediments in order of their crashing. Number 1 marks the Dahlonega Gold Belt, a narrow belt of mostly quartz muscovite schist (p. 105) that follows the red fault line across the State. Number 2 marks the Coweeta Pluton (now called the Persimmon Creek Gneiss of the Coweeta Group of the Cowrock Terrane, whew!) where it came to rest partially on top of the Dahlonega Gold Belt. The Coweeta Group is colored purple on the map and consists mostly of biotite gneiss (p. 115). Number 3 marks sediments (green on map) caught between Laurentia and the volcanic island . These consist of quartzite (p. 100), metagraywacke (p. 88) and muscovite schist (p. 108). Finally, number 4 marks the volcanic island (now called the Inner Piedmont) that came crashing last in the sequence.

The light pink color on the geologic map shows granite plutons emplaced when Africa crashed. These include Stone Mountian, Panola Mountain, Arabia Mountain, Elberton granite, Palmetto granite and other plutons. These are igneous rocks.

Can you find the very thin brown lines oriented from northwest to southeast? These are diabase dikes emplaced in EBRIP when the supercontinent Pangea, which included EBRIP, rifted apart about 150 MYA. Diabase, an intrusive igneous rock, made of dark (mafic) magma.

With the exception of granite plutons and diabase dikes, all other EBRIP rocks are metamorphic rocks. In fact, EBRIP is the metamorphic rock capital of Georgia!

The purple map colors are mainly metamorphosed plutons. Light purple is granite gneiss (light colored felsic magma), emplaced as granite when EBRIP crashed into Laurentia. Medium purple is biotite hornblende gneiss (intermediate magma), emplaced as diorite via fractional melting during EBRIP's volcanic island phase, and the dark purple is amphibolite (dark magma) emplaced as gabbro during frational melting The purple shades could also represent lava (above ground) flows of these three rock types.

The gray map colors represent primarily muscovite schist and sillimanite (very highly metamorphosed) schist, pyroclastic rocks blown from EBRIP's volcanos, and subsequently metamorphosed via hydrothermal alteration.

Look for the small red dots scattered on the map. These are outcrops of ultramafic rock like serpentinite, soapstone and asbestos. They crop out along the Hayesville Fault where ultramafic ocean crust was caught up and smashed between EBRIP and Laurentia. The largest red spot is Soapstone Ridge in Dekalb county (south Atlanta), the remains of ocean crustsqueezed up between island arc segments as EBRIP crashed into Laurentia.

As previously mentioned, the crashing of Africa during the Alleghany orogeny (300 MYA) was massive in terms of the heat and pressure it applied to the rocks it was thrust upon. EBRIP's rocks were affected more than those of any other Georgia geologic province. Virtually all of EBRIP's rocks were folded, refolded, faulted (broken) and turned into metamorphic rocks.

As a result, the rocks of EBRIP are arranged in longitudinal stripes or bands across the land, with a different rock type in each band. The bands are oriented from northeast to southwest. The Appalachian Mountain range, stretching along the Atlantic coast graphically illustrates this feature. How did these rock bands form?

The following images show how the rock bands came about from the forces of metamorphism. The four lines of color represent four EBRIP rock types arranged in layers.

Pressure directed from southeast to northwest was applied to EBRIP rocks during its three episodes of collision. The arrows in the drawing represent this 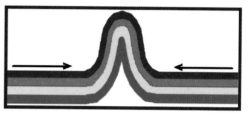 pressure (right arrow) and the corresponding resistance (left arrow). First, the rock layers were pushed into a huge upward fold called an anticline. There were really many anticlines, each separated by a corresponding dip in the layers called a syncline.

Continuing pressure pushed the anticline over on its side toward the northwest, to form a recumbent F1 (folded once) fold.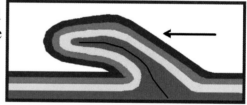

Still more pressure pushed the F1 fold into another anticline, which geologists call an F2 fold.

Finally, erosion washed away the top part of the F2 fold leaving a repeating series of rock stripes.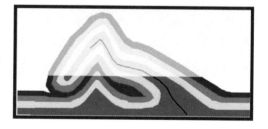

CHAPTER 7. CAROLINA TERRANE (CT)

The **Carolina Terrane** occupies the lower Piedmont province of Georgia. Its upper boundary is marked by the Middleton-Lowndesville fault, running about southwest from Lake Richard Russell. The Fall Line, running through Augusta, Macon and Columbus, marks its lower boundary, separating it from the Coastal Plain province to the south.

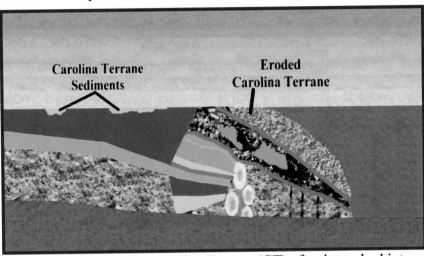

This picture shows the Carolina Terrane (CT) after it crashed into Laurentia plus EBRIP and then eroded down to sea level. The red line below CT is the Middleton-Lowndesville fault. CT left few sediments behind, probably because its "sidelong" collision with Laurentia created neither huge mountains nor an extensive forebasin to accept its erosion sediments.

CT's crashing occurred around 400 MYA and is called the Acadian orogeny. Incidentally, the word terrane is a term geologists use to refer to a hunk of land that experienced a unique geologic history. For example, EBRIP has been called the Tugaloo Terrane.

In the last chapter we learned that EBRIP began as a hunk of Laurentia that rifted off and moved only a short distance away. Subduction turned it into a volcanic island chain.

The Carolina Terrane also became a volcanic island chain, but geologists believe its history was different from EBRIP 's in several important ways.

First, unlike EBRIP, the CT island chain did **not** form around a hunk of rifted Laurentian crust. Instead, one piece of ocean crust was subducted beneath another, to create the CT volcanic island.

Second, the CT island chain formed close to Africa and far away from Laurentia. Geologistsbelieve this because they have found the same fossils in the CT as in Africa.

Third, EBRIP crashed "head on" into Laurentia, creating huge mountains that deposited thick sediments. In contrast, CT struck Laurentia in a kind of "side swipe". This sliding action did attach CT to the mainland, but it did not result in high mountains or massive erosion sediments as happened with EBRIP.

CT's sliding attachment did, however, cause a huge stress crack in EBRIP, with one side of the crack sliding along against the other. We know this giant stress crack as the Brevard Fault Zone.

Finally, Carolina Terrane rocks are very similar to EBRIP rocks, because they formed by volcanic island processes, but they have experienced less metamorphism than those of EBRIP. EBRIP received hard metamorphic knocks from both CT and Africa. CT endured only Africa's crash. This made it a little easier for geologists to figure out how CT rocks formed. Geologists learned a lot about EBRIP's origin by studying CT rocks.

After the Carolina Terrane eroded down to sea level, Pangea (the supercontinent formed by all the collisions we've been talking about) eventually rifted apart around 150 MYA. Much of the CT sank below sea level, and now underlies the Coastal Plain province as shown in the picture below.

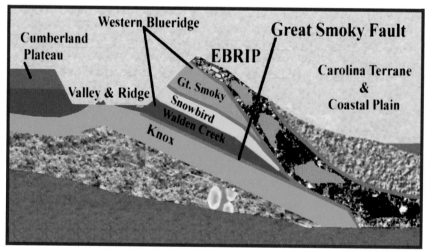

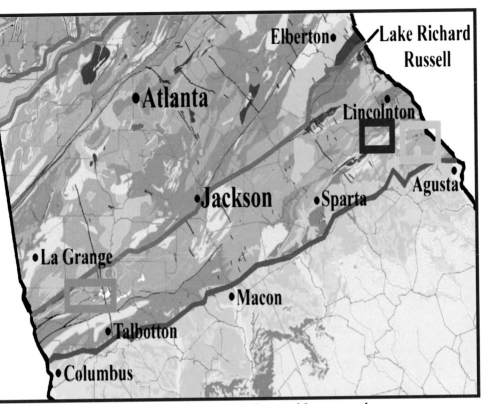

This map shows the Carolina Terrane located between the green line (Fall Line), and the red line above it (Middleton-Lowndesville fault). The Middleton-Lowndesville fault marks the place where the CT island chain crashed into Laurentia plus EBRIP during the Acadian orogeny. The Fall Line, where the Coastal Plain begins, is not a fault zone. The CT actually extends below the Coastal Plain sediments as the now buried rifted margin of Pangea

You can see by the map colors that CT rocks are very similar to those of EBRIP, but there are some interesting exceptions. Notice the dark pink areas in the northeastern portion of the CT. These areas, including towns like Rayle, Tignal and Lincolnton, represent outcrops of metamorphic felsic volcanic rocks.

Metadacite (p. 110) is one of these rock types. It is felsic lava that flowed stifly from the top of the CT island volcanoes. You will also find quartz muscovite schist here, the metamorphosed version of felsic tuff that exploded from the volcanos as pyroclastic rock and settled on its flanks. This unique area abundant in felsic volcanic rock is called the Lincolnton volcanic center by geologists.

The map orange rectangle marks the Warm Springs area, noted for 88-degree F. spring water emerging from the base of Pine Mountain. The mountain extends about 30 east-west miles through the area, and is made of WBR quartzite (Chilhowee Group.) deposited when Rodinia rifted (p. 12). Subsequent continental crashing literally scraped off overlying EBRIP rocks, exposing the quartzite, and then shoving it on its side, to create the mountain. Rain falling on the mountain descends 3000 feet through seems in the quartzite's edge where it heats up and reemerges as FDR's famous Warm Springs. The biologically diverse Sprewell Bluffs located where Pine Mountain intersects the Flint River is the same formation, as is (likely) Indian Springs State Park near Jackson, Georgia.

The blue rectangle on the map marks Graves Mountain (http://www.gamineral.org/graves_mtn.htm) near Lincolnton, the site of ancient island arc volcanic vents associated with CT's formation. Super hot water brought up minerals like titanium, iron, kyanite and lazulite from deep within the volcanoes, depositing them in host rock called pyritiferous kyanite granofels.

You already know that red circles on these geologic maps mean ultramafic rock. Notice the concentration of red circles in the upper right part of the map near the fault line. Geologists call this area the Russell Lake Allochthon. The allochthon includes ultramafic rocks like serpentinite and metapyroxenite, former segments of ocean crust squeezed up in the suture zone and spread across the land.

The yellow rectangle on the map northwest of Augusta surrounds some red ultramafic rocks on Burks Mountain. The mountain is the remnant of ocean crust squeezed up between the main bulk of the CT and an associated volcanic island segment called the Savannah River Terrane. Burks Mountain's ultramafic rocks have given rise to a special plant community called a serpentine barren. The high magnesium content of rocks like serpentinite actually acts as a soil poison, eliminating many of the plants and animals normally found in surrounding areas. On the other hand, plants adapted to dry nutrient poor soils and high concentrations of magnesium can grow here, unhindered by competition from the more typical plants.

CHAPTER 8. VALLEY & RIDGE PROVINCE
(V&R)

The **Valley and Ridge province**, located in northwest Georgia, lies between the Western Blue Ridge on the east and the Cumberland Plateau on the west. Today it consists of two large valleys separated by a line of low ridges. The Great Valley on its eastern side, is traversed by Georgia Highway 411 and Interstate 75, connecting towns like Chatsworth, Dalton and Ringold. The Chickamauga Valley runs along the western side of the province and includes towns like Chickamauga, Lafayette and Summerville.

The two valleys are separated by the Armuchee Ridges. Taylor Ridge, the longest, runs across the whole Valley and Ridge province. Shorter ridges spurring off of Taylor Ridge include Johns and Horn Ridges.

This familiar picture shows the rifted margin of Rodinia as it appeared around 550 MYA. Notice the pink layer of limestone sediments labeled Knox covering much of the WBR and all of the future V&R and Cumberland Plateau (CUP). These sediments were deposited as the rifted edge of Rodinia slowly sank below sea level in warm tropical waters, and the Cambrian explosion of life occurred (530MYA). We already know this pink layer actually consists of several different layers of rock topped by a thick layer of Knox limestone. The table on the next page lists these sediment layers (now hardened into rock) with the oldest sediments at the bottom.

Sediment/Rock Laye	Composition
Knox Limestone	Limestone
Conasauga Group	Shale and limestone
Rome Formation	Dusky-red and green shale and siltstone, dolomite, and limestone.
Shady Dolomite	Dolomite (limestone with magnesium added)
Chilhowee Group	Quartzite, conglomerate, sandstone and phyllite

When the edge of Rodinia began to sink, <u>river deltas, sand dunes and beaches</u> covered the land. These became the Chilhowee Group. Then the continent edge dipped below sea level. First, the Shady dolomite formed in hot shallow water, then <u>shales and limestones</u> of the Rome and Conasauga sediments formed in shallow offshore mud flats as the ocean level fluctuated back and forth. Finally, the thick limestone layer of the Knox Group was deposited as the land dipped deeper and remained below sea level.

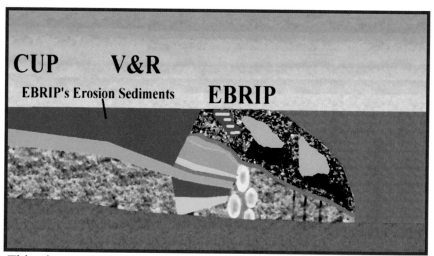

This picture shows the result of the Taconic orogeny that came next. EBRIP has crashed into Laurentia and subsequently eroded down to sea level. Its sediments (brown) have filled a foreland basin that covers what will become the V&R and CUP provinces.

You can see EBRIP's brown sediments covered all the pink rifting layers described in the table above.

Now let's take a closer look at EBRIP's sediments (brown layer). The oldest layers are at the bottom in the table below.

Sediment Layer	Composition
Red Mountain	Dark Red Sandstone
Sequatchie	Shale and Shaly Limestone
Chickamauga/Greensport	Limestone/Shale

The sediment sequence in the table shows us how these sediments formed. When the Taconic (EBRIP) mountains first began eroding, the foreland basin was fairly deep and far offshore from the eroding mountains. Limestone of the Chickamauga group formed. As erosion accelerated, the mountains got both <u>shorter</u> and <u>wider</u>, expanding westward into the basin. The deep water limestone basin became a shallow offshore marsh where mud deposits accumulated and eventually became the shale and shaly limestone of the Sequatchie Formation. With continued erosion and westward expansion, river delta and beach sands encroached on top of the shales. These became the sandstones of the Red Mountain formation.

In the V&R, the the Red Mountain formation consists of iron-rich red sandstone. It is easily recognized in the field by its intense dark red color (p. 86). Further west and north, a thick layer of very iron-rich hematite underlies the Red Mountain sandstone. The hematite formed by precipitation of abundant iron from sea water into sand and clay deposits and subsequent oxidation of the iron to hematite.

The Red Mountain hematite and Chickamauga limestone provided the two main ingredients for steel manufacture, and coal to power the iron furnaces was abundant in the nearby Cumberland Plateau. Birmingham, Alabama has flourished as a center of steel production based on these very sediments.

In the next act of our tectonic drama, Africa has crashed into Laurentia, pushing the edge of the continent down to form a foreland basin at the left in the picture. At this stage of the game, some of Africa's sediments (gray color) have already been deposited on the Valley & Ridge and the CUP. These sediments were deposited early on, when the foreland basin covered both CUP and V&R, and Africa was still being shoved up on the mainland. They consist of chert (Ft. Payne chert, p. 92) deposited in the CUP and western V&R where the basin was fairly deep, and shale (Floyd shale) deposited in the eastern V&R where offshore marshes were receiving fine-grained sediments eroded from Africa and washed into shallow water just offshore.

The red lines in the V&R represent faults where the pressure from Africa's collision pushed the rock layers into folds and then broke the folds, shoving them on top of each other. This probably raised the V&R above basin level as shown. Most of Africa's remaining sediments will wash into the extensive CUP basin (much larger than shown), eventually filling it up.

Though not graphically shown in the picture, all the rocks Africa touched experienced severe folding and faulting. The rocks of WBR, EBRIP and CT were deformed by this stress into metamorphic rocks, but V&R and CUP rocks, though folded and faulted, did not experience enough stress to produce metamorphic rocks. They remain as **sedimentary** rocks today.

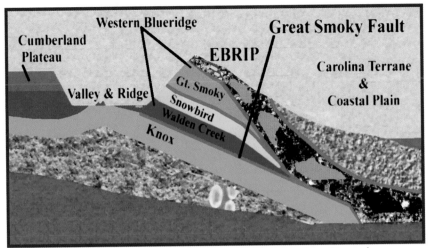

This picture shows the cross section of present-day Georgia. Africa rifted away around 150 MYA with the breaking up of the supercontinent Pangea. Most of its sediments (gray colors) eroded away too, except for a big stack in the CUP and a few traces in the Valley and Ridge.

The two brown spikes in the V&R represent the Armuchee Ridges, formed from erosion resistant Red Mountain sandstone. The Great Valley is on the right, the Chickamauga valley is on the left and the CUP is far left. Highway 136 (Lookout Mountain Senic Highway) between Resaca and Lafayette crosses the Armuchee Ridges and the Armuchee Valley. Red Mountain sandstone is exposed there in road cuts at the Pinhoti Trail parking lot.
After Africa's huge weight was removed, the continent slowly rose like a ship relieved of its cargo. Subsequent erosion defined the mountains and valleys we know today.

Incidentally, the Valley and Ridge province runs from Alabama up into Canada, providing easy north-south access across the eastern United States. From prehistoric times people have used this avenue for migration and trade, flourishing on crops from the rich limestone based soils. The Etowah Indian Mounds near Cartersville are the vestiges of a great Mississippian Period religious and cultural center, active between 1000 and 1500 A.D. Rich corn harvests supported the large population living near the mounds. Later, Cherokee Indians established many villages, including their capitol city, New Echota near present-day Calhoun.

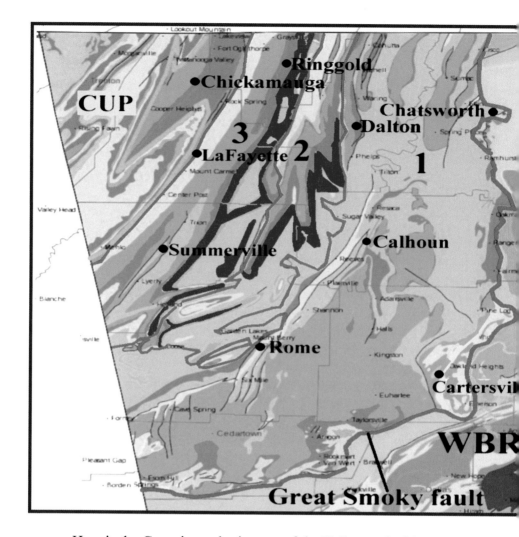

Here is the Georgia geologic map of the Valley and Ridge province showing the rock types found there today. The lower red line marks the Great Smoky fault separating the WBR on the right from the V&R. The yellow line is not a fault. It marks the cliffs of Lookout Mountain that form the boundary between the CUP on the left and the V&R.

The three lighter red lines mark faults where large folds, pushed up by Africa's crash were broken and then, one fold shoved on top of another. These faults have names, and from right to left they are the Rome fault, the Whiteoak Mountain fault and the Kingston fault. The rock sections between the faults are what's left of pieces of the great folds (anticlines and synclines) that once existed. Let's look closer at section 1 between the bold red Great Smoky fault and the light red Rome fault to the northwest. The light blue color in this section designates Knox limestone (USGS uses blue for limestone instead of pink). The dark blue color marks the older Conasauga limestone, and the green color marks the still older Conasauga shale. Remember these layers accumulated on the edge of Rodinia after it rifted apart and slowly sank below the sea.

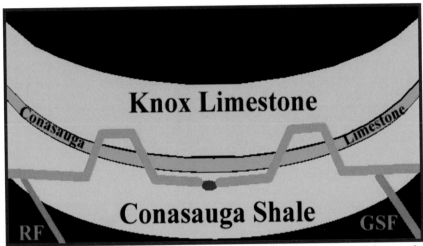

This picture shows generally how these rocks are arranged beneath much of the Great valley today. They form the remains of a downward bulging fold in the rock layers called a syncline. The brown line in the picture shows current ground level (vertical scale is exaggerated), and the dark blue spot represents the Conasauga River which has eroded through the Knox limestone and the Conasauga limestone, and currently flows in a bed of Conasauga shale. The two faults that bound the formation are drawn and labeled in red.

Incidentally, the Great Valley town of Cartersville Georgia is home to the Tellus Science Museum (http://www.tellusmuseum.org/), that houses a huge fossil display and the **Weinman Mineral Gallery,** the largest rock and mineral collection in the Southeast. Weinman pioneered the mining of barite, a mineral associated with limestone, in Georgia. He donated rocks, minerals and money to start the Weinman Mineral Museum.

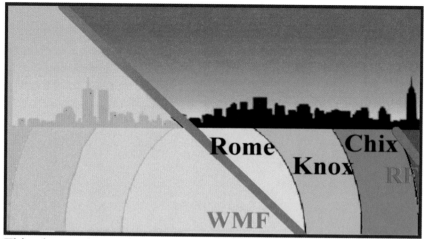

This picture shows the rock layers between the Rome fault and the Whiteoak Mountain fault. This is the section labeled **2** on the geologic map. The layers are the faulted and eroded remains of the right arm of an <u>anticline</u>, an upward bulging fold in the rock layers.

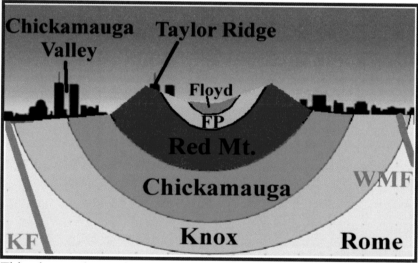

This picture shows the rock layers composing a large section of Taylor Ridge, the most prominent of the Armuchee Ridges (section **3** on map). The rock layers are part of a syncline that has been cut off by faults (Kingston and Whiteoak Mountain faults) on each side and then eroded. The Armuchee Valley lies below the word Floyd and is, in fact, made of the Floyd shale formation. The letters "FP" refer to the thin layer of Fort Payne chert, that consists of "limy" chert containing chert nodules that Indians once chipped into beautiful arrowheads, scrapers and spear points.
The left side of the picture above shows the rock layers of the

Chickamauga Valley just west of Taylor Ridge. They are the eroded remains of the left side of the adjacent syncline.

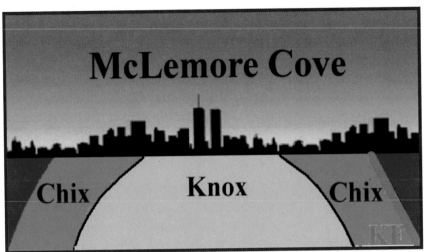

This picture shows the rocks of McLemore Cove, located west of the Chickamauga Valley between Pigeon Mountain and Lookout Mountain. Surrounded by high and picturesque cliffs, this beautiful valley is devoted almost exclusively to farming, a testament to the rich and abundant limestone soils. Eastern red-cedar, a calcium loving tree is very abundant here. The cove rests on an eroded anticline exposing Chickamauga (Chix) and Knox limestone formations.

You have probably noticed that anticlines often form valleys when they erode, and synclines often form ridges. This seems opposite of what you'd expect, but it really makes sense. An anticline (upward bulging fold) develops many cracks when the rock layers are folded upwards that readily allow water to penetrate the rock layers, and so they erode relatively quickly down to valleys like the Chickamauga Valley and McLemore Cove.

On the other hand, the rock layers in a syncline are pressed together very tightly. It takes water a long time to penetrate the layers, so they end up as ridges like Taylor Ridge, and the other Armuchee Ridges.

CHAPTER 9. CUMBERLAND PLATEAU (CUP)

The Georgia **Cumberland Plateau** lies on the western flank of the Valley and Ridge province in the form of Pigeon, Lookout and Sand Mountains. At about 1800 feet above sea level, the Cumberland Plateau rises 800 feet above the adjacent Chickamauga Valley in a series of sheer sandstone cliffs separated by slanting slopes of shale. Let's dive into its geologic history.

This picture from the last chapter shows the CUP on the far left. Africa has recently (300 MYA) crashed into Laurentia, forming the supercontinent Pangea. At this point in Africa's erosion process, limestone (shone as gray color here) has formed in the CUP foreland basin (far left), but Africa still has a lot of sediments to shed as it erodes down to sea level. They will accumulate in this basin that extends many miles across the Pangean continent.

Africa will continue to get shorter and wider as it erodes. As its sediments continue to push west, mud (future shale) will be dumped on top of the limestone, and then sand (future sandstone) will be dumped on top of the mud. In other words, what once was a fairly deep basin will first become salt marsh and mud flats (shale), and finally a flat sandy coastal plain (sandstone).

If you looked at a road cut showing rocks associated with this process where the land is **encroaching on the sea** (foreland basin), you would see a layer of limestone (Mississippian Limestone), then a layer of shale and finally a layer of sandstone and conglomerate marking the final emergence of land above sea level.

If the land encroached in fits and starts (as it did in the CUP), there would be alternating layers of mud (flats and marshes) and sand (deltas and beaches) with a final layer of sand marking the permanent emergence of dry land.

This picture shows the Cumberland Plateau after the sand moved in, turning Georgia's portion of the CUP into a coastal plain. Africa's sediments are still flowing into the basin at left which is far larger than shown. Basin sediments will eventually spread over Tennessee, Kentucky, Ohio and beyond.

For the first time ever, massive swamp forests of tree ferns and other primitive fern-like plants line the great coastal plain rivers that dump Africa's sediments into the CUP basin. These forests are massive in scope, fringing the shores of Africa's giant foreland basin for thousands of miles. As plants in this forest die, their remains fall into the swamp and become peat. This same process is happening today in the Okeefenokee Swamp.

The primitive forests and peat swamps last for millions of years, while great ice sheets come and go many times. Each time an ice sheet melts, sea level rises, and the peat swamps are covered by a layer of mud that will eventually turn to shale. When the sea recedes, a layer of sand will cover the shale, and swamp forest will return again.

The many buried layers of peat will eventually be turned into rich coal deposits. When Pangea breaks up (150 MYA) these coal deposits will be spread throughout the world, forming most of the world's great coal reserves.

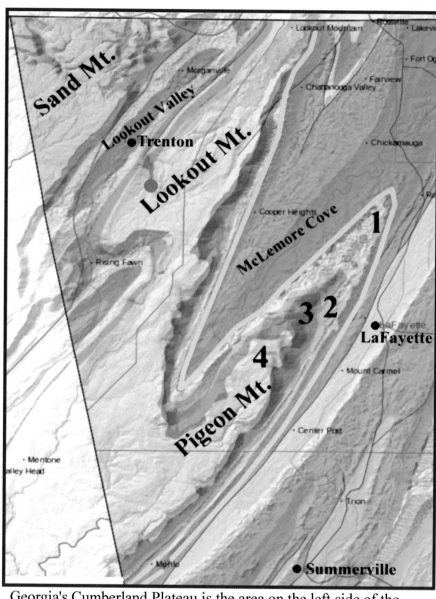

Georgia's Cumberland Plateau is the area on the left side of the yellow line in this picture. The numbers superimposed on Pigeon Mountain mark the rock layers of the CUP. The greenish layer labeled number 1 is the iron rich Red Mountain sandstone (p. 86). This formation is marked in maroon on the V&R chapter map. It is the oldest rock layer visible in the CUP, representing the last of EBRIP's erosion sediments to be deposited.

Number 2 marks the Fort Payne chert (p. 92) (gray) that lies on top of the Red Mountain sandstone. Fort Payne chert is the very earliest of Africa's sediments, deposited in Africa's foreland basin when it was very shallow and a hot dry climate prevailed. Fort Payne chert also covers the western V&R.

Number 3 in the picture marks a formation called Mississippian limestone (p. 94). This is a generic term that really refers to several formations that include mostly limestone with some shale. These deposits were laid down when Africa first rode up on Laurentia, pushing Laurentia down and causing the foreland basin to become deep enough for a thick layer of limestone to form. This layer is absent from the V&R, probably because the V&R was folded, faulted and shoved up above basin level by Africa's crashing.

Number 4 in the picture represents Pennsylvanian sandstone (p, 86), consisting of several named layers of mostly sandstone and conglomerate with some shale mixed in and coal deposits at the highest levels. The Pennsylvanian sandstone capping the CUP, protected layers beneath it from eroding, resulting in the elevated plateau region we see today.

Coal deposits once topped the sandstone. Though most of Georgia's coal layers were eroded away, states like Pennsylvania, Ohio and West Virginia still have abundant rich deposits of coal derived from the ancient fern forests of Pennsylvanian time.

Cloudland Canyon State Park (red circle on the map) on Lookout Mountain is a great place to see the rock layers of the Cumberland plateau. The sides of the canyon are a series of sheer sandstone cliffs separated by steeply angled slopes of mixed shale and sandstone. These features are made of Pennsylvanian sandstone eroded from Africa. Trees like sand hickory that can handle dry, nutrient-poor sites grow on the sandstone in numbers not found elsewhere. American holly, a tree adapted to dry acid soils is also abundant here.

The canyon lower slopes and bottom are made of Mississippian limestone, a rock type producing fertile, calcium-rich soils where calcium-loving trees like sugar maple and shagbark hickory thrive.

Limestone is fairly easily eroded by rain water, a weak acid. The resistant Pennslyvanian sandstone above forms a roof over an extensive network of caves carved in the Mississippian limestone below. Pigeon Mountain is a nationally famous caving destination, providing caving adventure for underground enthusiasts, including Racoon Mountain Caverns, Ruby Falls and Ellison's Cave, the deepest cave east of the Mississippi River.

Georgia's CUP is also a good place to go fossil hunting. Lula Lake Land Trust Park, located on the northern point of Lookout Mountain is the former site of the Durham Coal Mine. Excellent examples of Pennsylvanian fern fossils can be found here, in piles of shale removed from the coal deposits during the strip mining process.

Interstate 24, in the extreme northern part of Georgia's CUP runs in road cuts that expose Mississippian limestone containing fossils from this period. Check the internet or your local rock club for more information about these CUP fossil sites.

Fern fossil in Pennsylvanian shale

CHAPTER 10. COASTAL PLAIN (CP)

The **Coastal Plain** covers the southern two thirds of Georgia. Its northern boundary, called the Fall Line, runs through Augusta, Macon and Columbus. The Coastal Plain is geologically young, consisting mostly of sand and clay sediments interspersed with limestone and scattered sandstone deposited during the last 150 million years. The imprint of the sea that once covered the region is everywhere. Let's take a close look at the origin of this land of sand and sea.

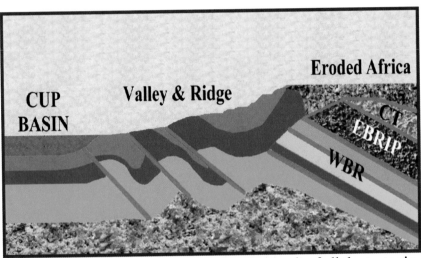

Here is a picture of Laurentia showing the result of all the tectonic events it experienced between about 1 billion years ago and 300 MYA when Africa crashed into it. The WBR sediments, deposited when Rodinia rifted, were covered by EBRIP when it was shoved on top of them during the Taconic orogeny (500 MYA). Then, EBRIP eroded down to sea level, depositing a layer of sediments (brown) in the V&R basin.

Next, Carolina Terrane "sideswiped" EBRIP and was sutured to it during the Acadian orogeny (400 MYA). It left few sediments to mark its erosive demise.

Africa was shoved on top of everything during the Allegheny orogeny about 300 MYA. Africa folded and faulted all previous rocks. It shoved WBR, EBRIP and Carolina Terrane miles on top of the Knox limestone and everything beneath it. Africa eventually eroded, depositing its sediments (gray) in the CUP basin.

All this crashing created the world's last great supercontinent, **Pangea**, which lasted from about 300 MYA until about 150 MYA, when it too rifted apart and the world's continents took on their approximate present configuration. Pangea's existence covered three great geologic time spans, the **Permian**, **Triassic** and **Jurassic** periods (p. 68). By the time these great periods were finished, Pangea was gone.

A glance at the previous picture shows the rocks that will become Georgia are landlocked by eroded Africa and residing slightly above sea level. Consequently, few sediments were deposited during these three great spans of time when Pangea existed. Instead, erosion occurred, so Georgia and most of the eastern U.S. have few rocks to document these periods. Dinosaurs and eventually mammals began roaming the earth, but their early stages are mainly documented in the Western States where sediments *were* accumulating.

Here is the close up edge of the Laurentian continent around 100 MYA, after Pangea has rifted and the growing Atlantic ocean separates Laurentia and Africa. The Carolina Terrane is now at the continent edge. A layer of sand (brown), probably deposited during the first stages of rifting, is being covered by mud (blue) and limestone (pink) as the rifted continent edge sinks. Sea level worldwide is also rising because the climate is warm, and little water is locked in polar ice and glaciers.

At the beginning of the Eocene epoch, around 65 MYA the massive meteor called Chixulub crashes into Earth at the Yucutan

66

Peninsula, causing catastrophic climate change, and bringing the dinosaur age to an abrupt end.

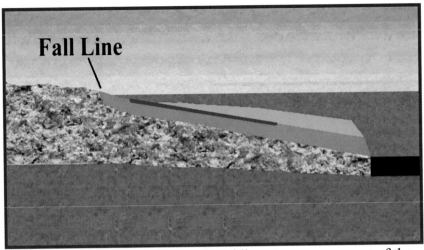

As the sea continues to rise, mud and limestone cover most of the sand layer. This occurs during the Eocene (65-35 MYA) and the Oligocene (35-25 MYA). The ocean rises to it greatest height ever in Georgia, marking our present-day Fall Line.

It is now the Neogene epoch (25-.05) MYA. The world's continents are in their present-day places, and polar ice caps are forming. This lowers sea levels, resulting in a new layer of sand and mud deposited on the Georgia Coastal Plain as sand beaches and mud flats "chase" the sea back down the slope. The limestone layer gets covered over.

Fall Line

Erosion has now occurred along the coast, exposing the three main sediment layers with the oldest layer (sand and mud) along the Fall Line, and the youngest layer (more sand and mud) along the coast. A layer of limestone (pink) lies between the two sandstone layers.

In the last act, ice age related sea level fluctuations shape the lower coast lands, creating interesting sand ridges and a huge swamp.

Table of Geologic Time

Quaternary (Pleistocene)	**1.5-.05 MYA**	Ice age fluctuations cause sea level fluctuations that shape and reshape lower Coastal Plain.
Tertiary (Neogene)	**25-1.5**	Polar icecaps are forming. Sea level is going down. Sand deposited over the limestone.
Tertiary (Oligocene)	**35-25**	Oceans continue to rise. More limestone deposited.
Tertiary (Paleocene/Eocene)	**65-35**	Chixulub meteor hits (65 MYA). Dinosaurs are wiped out! Limestone deposited over sand.
Cretaceous	**150-65**	Sand deposited on Coastal Plain. Pangea rifts apart, and oceans rise.
Jurassic	**200-150**	Dinosaurs rule!
Triassic	**250-200**	First dinosaurs and mammals.
Permian Period	**300-250 MYA**	Pangea forms when Africa crashes.

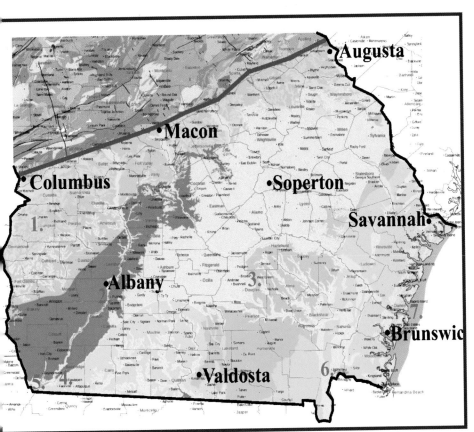

Here is the current geologic map of the Georgia Coastal Plain province. The tectonic forces just discussed have shaped the Coastal Plain (CP) into **6 regions**, visible as the various colors on the map. Let's look at each one.

The yellow region at the top is called the **Fall Line Sand Hills**, and consists of the oldest layer of sediments shown on page 66. These are the sands and muds deposited during the late Cretaceous and early Eocene, and they include the <u>Irwington Sand</u>, <u>Twiggs Clay</u>, <u>Clinchfield Sand</u>, <u>Providence Sand</u> and others.

The Fall Line (green line on map) represents the ocean's highest encroachment on the land during the last 50 million years. It is marked by a line of small water falls and cascades where Piedmont streams flow into the Coastal Plain. In the past, ships could not go beyond this point, so important "port" cities like Augusta, Macon and Columbus developed where produce and manufactured goods from the Piedmont were loaded on ships anchored just below the fall line. The abrupt change in stream elevation along the Fall Line also provided an outstanding place for water-powered factories that stimulated development of Fall Line towns.

Sixty five MYA ,when ocean waves reached this far, great swamps stretched for miles along the coast. Piedmont rivers dumped huge loads of eroded sediments into the swamps including great quantities of feldspar-rich clay. Over time, these deposits became kaolin (p. 93), a high quality pure white clay, of great commercial value today. Georgia is by far America's leading producer of kaolin which is used in the making of glossy paper, commercial ceramics and even tooth paste.

Another interesting geology-related feature of the Fall Line Sandhills is Providence Canyon State Park, in Stewart County (red 1 on map) where poor farming practices and erodible sands have created colorful canyons. The bright colors in the canyon walls result from oxidation of iron and manganese contained in the sands and clays. Sir Charles Lyell, the foremost geologist of the 19th century, wrote about the canyon in his monumental book, Principles of Geology, first published in 1830.

The blue map area, called the **Dougherty Plain**, consists of limestone. It is the middle (pink) layer in our earlier picture. The large blue area in the southwest is the Ocala Limestone (p. 96). To the northeast, the blue area becomes the Suwannee Limestone. Both limestones are part of the huge Floridian Aquifer that underlies south Georgia and Florida providing these areas with fresh water for city water supplies and rural crop irrigation.

Limestone is made of calcium carbonate, a base. Water, a weak acid, dissolves the limestone, forming extensive flooded caves that stretch for hundreds of miles below ground and can be tapped as a water source.

Where a limestone cave opens to the surface, deep freshwater springs often result. Radium Springs in Dougherty County near Albany (red 2 on map) is Georgia's largest limestone spring. A century ago folks flocked to its waters seeking recreation and health benefits. Originally called Blue Springs (calcium in the limestone tints the water blue), its name was changed when trace amounts of radium were detected in the water, a discovery that "enhanced" its so called health benefits.

A visit to one of several limestone quarries in the Dougherty Plain, treats the visitor to a graphic visual record of the area's geology, as well as a great fossil collecting opportunity. The thick limestone layer of the quarry is covered by a thinner layer of Twiggs clay, a silica rich light gray clay made of the microscopic bodies of tiny marine organisms called diatoms. Fuller's earth (p. 92), an older name for high-silica clay like the Twiggs clay, was once used by fullers (wool cleaners) to remove oil and dirt from newly woven wool fabric, a testimony to the clay's high absorption capacity.

The thick layer of Ocala limestone below the clay is a loose white jumble of calcium carbonate, scallop shells, sand dollars and other marine organisms. Pods of glassy reddish chert can also be found. These formed when water, circulating through the limestone picked up silicon from the clay and deposited it as chert in pockets in the limestone.

The light green area on the Georgia map marks a vast region of sand underlain by sandstone and claystone deposited during the Neogene epoch (25-.05 MYA). Called the **Altamaha or Tifton Uplands,** this wide formation, stretching across the whole state was once the heart of the noble Longleaf pine-wiregrass forest community.

Before Europeans arrived, fires started by lightening and Indians burned the high dry ridges here about every 2 or 3 years. Burning halted the normal progression of forest succession, excluding fire intolerant hardwood trees and favoring the fire resistant longleaf pine (*Pinus palustris*) and wiregrass (*Aristida beyrichiana)*. This created what ecologists call an edaphic pine climax forest, where frequent burning maintains pine as the climax tree. Early observers of the longleaf pine-wiregrass forest marvelled at its open park-like appearance. Game animals like deer, turkey, bear, squirrel, and quail thrive in such an environment.

Another interesting natural feature of the Altamaha Uplands is Broxton Rocks (red 3 on map), a large sandstone outcrop along a 4 mile stretch of Rocky Creek in Coffee county. This unique habitat, part of the Altamaha Grit formation, supports over 500 plant species, including Appalachian mountain, coastal plain and topical plants, many of which are rare and endangered.

The medium yellow zone covering extreme southwest and south central Georgia is called the **Limesink Region**. Here, a thick layer of sand and mud lie on top of the Ocala limestone (p. 96). The sand and mud erode, creating steep ravines and exposing openings in the limestone that connect to the extensive network of caves and springs associated with the Floridian Aquifer. Decatur and Grady counties are host to several notable ravines and sinks such as Climax Sink and Caverns near the small town of Climax, Georgia, Glory Caverns in Grady County and Waterfall Sink, also in Grady County (red 4 on map).

In extreme southwest Georgia (red 5 on map), along the Flint River arm of lake Seminole, the limesinks of this region have produced steep cool ravines where trees like the very rare Georgia native evergreen Torreya (*Torreya taxifolia)* are struggling for survival. These cool ravines provided a place of refuge for many plants when ice age glaciers swept down from the north as late as 10,000 years ago. When the ice receded, this association of cool climate trees and herbs was trapped in the cool ravines of the Limesink Region. Mountain trees like basswood and umbrella magnolia grow alongside torreya, southern sugar maple and mountain wildflowers like bloodroot (Sanquinaria canadensis) and dutchman's pipe (*Aristolochia tomentosa).*

The dark yellow zone southeast of the Altamaha Uplands is called the **Pliocene Sand and Gravel Region**. It consists of the sand of former beach and dune deposits left behind as the ocean receded for the last time, so far! The Okefenokee Swamp (red 6 on map), a former marsh trapped by dunes when the ocean receded, is located here. Trail Ridge which forms the swamp's eastern border was once a long line of barrier dunes stretching deep into Florida. The ridge continues to impound this, the largest peat swamp in the United States (430,000 acres).

Finally, the light yellow terrain along the coast marks the youngest geologic deposit in Georgia, consisting of sands and clay of the **Shoreline Complex**. This area is often referred to as the Georgia flatwoods, just a few feet above sea level, and only recently above its influence.

Georgia's "Golden Isles" (red 7 on map) provide some of the State's best beaches and ocean fishing, as well as nesting habitat for loggerhead sea turtles and extensive salt marshes, the nurseries of the sea. Islands like Jeckyl, Sapelo and Skidaway formed from sand brought to the coast by mighty rivers like the Savannah, Altamha and St. Mary's.

Sapelo Island, Georgia

CHAPTER 11. ROCK ID TABLES, DESCRIPTIONS & PHOTOS

To identify Georgia rocks, go to the appropriate province table in the pages that follow. CRACK YOUR ROCKS and put them in felsic, intermediate, mafic and ultramafic piles based on shade, color and texture. Now, check the appropriate table for a match and then look at the rock photos in the book.

For difficult rocks, use a hand lens to identify the rock's minerals and each mineral's relative abundance. Notice mineral grain size and whether the mineral grains are all the same size or of several sizes (turbidites).

WESTERN BLUE RIDGE (WBR) ROCKS

ROCK NAME	ROCK CLASS	TYPE	PAGE
Quartz Vein rock	Meta.	Felsic	98
Sandstone	Sed.	Felsic	85
Quartzite	Sed.	Felsic	100
Conglomerate & Turbidite Sandstone	Sed.	Felsic	83
Metagraywacke	Sed.	Felsic	88
Metasiltstone	Meta.	Felsic	103
Slate	Meta.	Felsic	103
Phyllite	Meta.	Felsic	103
Muscovite Schist	Meta.	Felsic	108
Granite Gneiss (augen)	Meta.	Felsic	112
Amphibolite	Meta.	Mafic	118
Marble	Meta.	--------	125
Ultramafic Rock	Meta.	UMafic	122

EBRIP AND CAROLINA TERRANE ROCKS

ROCK NAME	ROCK CLASS	TYPE	PAGE
Quartz Vein Rock	Meta.	Felsic	98
Quartzite	Meta.	Felsic	100
Granofels	Meta	Felsic	101
Metachert	Meta.	Felsic	102
Slate & Phyllite	Meta.	Felsic	103
Metagraywacke	Meta.	Felsic	88
Quartz Muscovite Schist	Meta.	Felsic	105
Muscovite Schist	Meta.	Felsic	108
Metadacite	Meta.	Felsic	110
Granite	Igneous	Felsic	77
Granite Gneiss	Meta.	Felsic	112
Biotite Hornblende Gneiss	Meta.	Intermediate	115
Amphibolite & Metagabbro	Meta.	Mafic	118
Diabase	Igneous	Mafic	80
Weathered Mafic Rock	Meta.	Mafic	121
Serpentinite, Soapstone, Corundum	Meta.	Ultramafic	122

EBRIP and CT have only 2 igneous rocks: granite and diabase. All the rest are metamorphic.

Granite gneiss usually has metamorphic banding. Granite doesn't. Use the nickel scratch test (page 115) to distinguish biotite from hornblende. Test for iron content of a rock by putting powdered samples in hydrogen peroxide, where they should fizz. Soapstone (talc schist) can be scratched with a fingernail.

VALLEY & RIDGE (V&R) AND CUMBERLAND PLATEAU (CUP) ROCKS

ROCK NAME	ROCK CLASS	TYPE	PAGE
Conglomerate	Sed.	Felsic	83
Sandstone	Sed.	Felsic	85
Chert	Sed.	Felsic	91
Shale	Sed.	Felsic	89
Limestone	Sed.	---------	94
Dolomite	Sed.	---------	94
Coal	Sed.	---------	97

You can usually scratch shale with a fingernail. Fossils are often found in shale.

A steel knife will scratch calcite, the mineral in limestone, dolomite and marble, but it will not scratch quartz, chert or flint.

Test for calcium carbonate (calcite) in limestone, dolomite and marble by pouring vinegar over a powdered sample of the rock. Calcite will fizz.

COASTAL PLAIN (CP) ROCKS

ROCK NAME	ROCK CLASS	TYPE	PAGE
Sandstone	Sed.	Felsic	85
Chert	Sed.	Felsic	91
Shale	Sed.	Felsic	89
Limestone	Sed.	---------	94
Kaolin	Sed.	Felsic	93
Fullers Earth	Sed.	Felsic	92

Igneous Rocks

Granite

<u>Rock Class:</u> Felsic, Igneous
<u>Primary Minerals</u>: quartz, potassium feldspar, plagioclase feldspar
<u>Secondary Minerals:</u> biotite mica, occasionally muscovite mica
<u>Province:</u> EBRIP, CT

For a rock to be called granite, it must have quartz (looks gray or clear) and feldspar (looks white or ypinkish) as primary minerals. The feldspar is often potassium feldspar, but it can also be plagioclase (sodium-rich) feldspar or a mix of the two.

The feldspar in granite may be present as large white or pinkish crystals (phenocrysts) surrounded by smaller quartz crystals like that of the Siloam, Georgia granite. More commonly, though, it lacks phenocrysts. Biotite (black specks) mica is usually present in granite as an accessory mineral, with rare muscovite (silvery).

Distinguish granite from granite gneiss (its metamorphic equivalent) by its fresher, less weathered look and lack of metamorphic banding. In Ggeorgia, granite is far rarer than granite gneiss.

Granite is a felsic intrusive igneous rock that forms when felsic magma squirts up into deep cracks and pockets in other rock. This happens during continental rifting, and during island arc and continent collisions. The magma that forms granite never reaches the earth's surface, but later, after many, many years it is exposed by erosion of the overlaying rock.

The Georgia EBRIP hosts two well known granite formations. Stone Mountain in Dekalb County, just east of Atlanta is a granite pluton that extends about nine miles underground into adjacent Gwinette County. It's gray in color and contains equal parts of potassium feldspar and plagioclase feldspar with significant quartz and accessory biotite and muscovite,

The much larger Elberton granite batholith (means a really big hunk) in Elbert County is <u>blue-gray</u> in color, and composed of quartz, potassium feldspar, plagioclase feldspar, biotite and minor muscovite. It is outstanding for monuments and buildings.

Other Georgia granites include Panola Mountain, Siloam, Sparta, Appling and Danburg. These granites intruded when proto-Africa crashed into proto-North America about 300 MYA during the Alleghany orogeny. Since that time, no tectonic event capable of metamorphic deformation has occurred, so they remain as granite instead of granite gneiss.

Elberton granite, unweathered from the quarry

Siloam granite with big potassium feldspar crystals

Stone Mountain granite

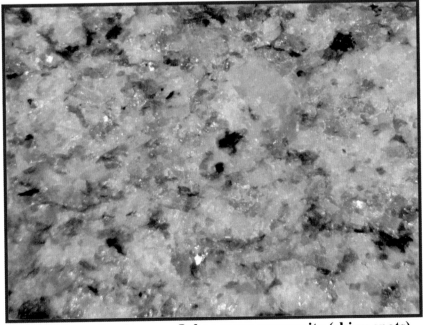

Stone Mt. granite closeup. It has more muscovite (shiny spots) than Elberton granite.

Diabase

<u>Rock Class:</u> Mafic (slightly more than gabbro), Igneous
<u>Primary Minerals:</u> plagioclase feldspar (calcium-rich), pyroxene, little or no quartz, no biotite
<u>Secondary Minerals:</u> hornblende, minor olivine
<u>Province:</u> EBRIP, CT,

Diabase looks dark gray, dark bluish or black and usually fine to med-grained,. It can have a crystalline look and feel and a rounded weathering pattern, both also characteristics of amphibolite. The weathering rind coating a piece of diabase and the soil produced by this rock are tan or brownish gray in color, instead of the typical red of Georgia soils. A good hand lens may show the <u>lath-shaped</u> (relatively long, ragged edged) crystals of plagioclase feldspar. They may also look <u>needle-like</u> and bunched. They look transparent instead of white.

Distinguish diabase from amphibolite by its lath and/or needle-like plagioclase crystals, its subtle dark bluish tinge (locals sometimes call it bluestone) and by its location <u>only where a diabase dike is indicated</u> on the Georgia Geology map (http://mrdata.usgs.gov/sgmc/ga.html).

Diabase dikes are long (several miles), but quite narrow (usually less than 1000' wide) bodies of diabase, most likely emplaced when Pangea (the supercontinent formed by Laurentia plus EBRIP plus Carolina Terrane plus proto-Africa) rifted for the final time, around 150 MYA.

The hot spot that did the rifting "boiled" up beneath Pangea, thrusting feeder dikes of diabase magma out in a radial pattern from its center. These monster blow torches melted the continent in two. Since then, no collision event has occurred to metamorphose this rock.

Diabase dikes were also formed when Rodinia first rifted (750-550 MYA), but these dikes have most likely been metamorphosed into amphibolite by all the continental pushing and shoving subsequent to their emplacement. Their more deformed and highly weathered state makes them difficult to locate as dike outcrops.

Diabase with tan weathering rind

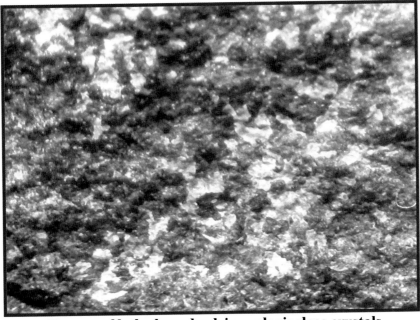

Closeup of lath-shaped calcium plagioclase crystals

Coarse grained diabase with needle-shaped plagioclase crystals

Closeup of needle-shaped plagioclase crystals

Sedimentary Rocks

Conglomerate and Turbidite Sandstone

Rock Class: Felsic, Sedimentary
Primary Minerals: quartz
Secondary Minerals: feldspar
Province: WBR (metaconglomerate), CUP

Conglomerate is sandstone with pebbles embedded in it. The pebbles may be rounded, showing they have traveled long and far in streams, or they may be angular, suggesting they have experienced less stream time and distance. The pebbles range from ant size to robin's eggs and larger.

Conglomerate usually forms from sand and pebbles in a stream bottom. The heavy pebbles sink to the bottom in a sand matrix where they harden into conglomerate. When you look at rocks formed in a river bed, you usually see a sandstone layer on top with conglomerate below, marking the river bottom.

Turbidite sandstone is also called both conglomerate and sandstone by geologists. It forms as sediments are rapidly deposited in a deep, steep-sided graben or basin where sand and rough-edged pebbles tumble down together and get mixed up size-wise.

Geologist believe WBR turbidite sandstone (they call it Thunderhead sandstone) formed when a new horst split the WBR graben associated with Rodinia's rifting (page 31). Material eroding from the horst washed quickly into the basin, tumbling down its steep sides and getting jumbled up size-wise in the process.

The first photo shows Lookout Mountain conglomerate from the Cumberland Plateau (CUP). These smooth rounded pebbles tell the story of a thousand-mile or more journey down an Amazon-size river flowing along the flanks of the mighty mountains created when Africa crashed into the continent during the Alleghany orogeny.

The second photo below shows WBR turbidite sandstone with small pieces of blue quartz. The blue quartz is characteristic of much of the WBR turbidite sandstone, having eroded from ancient augen gneiss granite in the new horst that split the WBR basin in two.

Lookout Mountain (CUP) conglomerate

Sandstone

Rock Class: Felsic, Sedimentary
Primary Minerals: quartz
Secondary Minerals: hematite iron
Province: WBR (metasandstone), V&R, CUP, CP

Sandstone is made of sand grains pressed into a rock. It comes in a variety of colors including red, tan, brown, gray and white. If you crack sandstone, the crack develops between the sand grains. If you crack quartzite (sandstone's metamorphic equivalent), the crack often develops through the grains.

The first photo below is Lookout Mountain (CUP) sandstone. There, it forms three rock cliffs separated by sloping shale layers. CUP sandstone marks former coastal plain rivers, river deltas and beaches associated with erosion of mountains formed during the Alleghany orogeny. A layer of conglomerate at the base of a big solid sandstone cliff tells you this was once a river bed. If instead, you find more narrow multi-layered beds of this sandstone and can see ripple marks on some of it, then you are looking at old sandy tidal flats once near the sea.

The second photo below is Red Mountain sandstone from the V&R. This specimen came from the Pinhoti Trail head on Highway 136 where it crosses the Armuchee Ridges. The red color comes from abundant iron that gave rise to early iron foundry towns like Birmingham, Alabama. This sandstone marks the latter stages of erosion of the Taconic Mountains (EBRIP) into the V&R foreland basin (after 500MYA).

The third photos shows fine grained gray sandstone from the WBR. Unlike graywacke, this sandstone did not form as a turbidite sediment. Instead, it was likely laid down on the ocean side (deeper water) of a river delta. The small consistent grain size gives this rock a kind of solid crystalline feel similar to amphibolite. It often weathers into rounded shapes.

The last photo is Altamaha Grit sandstone from the Altamaha Uplands region of the Coastal Plain. It formed when glaciers locked up ocean water in polar ice caps around 20 MYA. Sea level dropped and a layer of sand was deposited as the beach "chased" the ocean back down the slope.

Lookout Mountain (CUP) sandstone

Red Mountain sandstone

Fine grained WBR sandstone

Altamaha Grit sandstone

Graywacke (metagraywacke)
Rock Class: Felsic, Sedimentary
Primary Minerals: quartz
Secondary Minerals: feldspar
Province: EBRIP, CT, WBR

Georgia graywacke is really metagraywacke since it has experienced some metamorphism. It is light to dark gray in color and made of **variable size grains** of quartz and feldspar imbedded in a fine-grained matrix. You can consider it a finer grained version of turbidite sandstone (p. 83), and it formed the same way by rolling down steep graben sides in underwater avalanches.

To identify this rock look for variable size grains (use a hand lens). That is the key. If the largest hunks of quartz and feldspar are less than about 1/8-inch, then you likely have metagraywacke instead of turbidite sandstone. The only physical difference between this rock and fine grained WBR sandstone is its **variable** grain size.

The photo is from the Copperhill Formation of the Great Smoky Group (WBR), and represents sediments washed into the East graben basin after the new horst split the first basin in two (p. 31).

Shale

Rock Class: Felsic, Sedimentary
Primary Minerals: quartz, feldspar
Secondary Minerals: carbon from decomposed plants and
sometimes calcium carbonate from sea creatures
Province: WBR, V&R, CUP

Shale can be gray, black, tan and other colors. It forms platy layers
that you can often break apart with your fingers. It has a dull (not
silky or shiny) luster. Shale forms when mud settles in water and
is pressed into a soft rock. It is a great rock for preserving fossils.

The first photo below is Conasauga shale from Highway 411 south
of Chatsworth in the Great Valley. This shale was deposited in
shallow mud flats along the east coast of Rodinia as it sank into the
sea around 550 MYA after rifting apart. Eventually the land sank
further and the Conasauga limestone and the thick Knox limestone
formed above it.

The next two photos are Mississippian shale deposited in the CUP
as Africa's eroding sediments pushed west, covering the
Mississippian limestone with shallow mud flats just off shore.
The warm shallow water of the mud flats teemed with primitive
creatures like Crinoids and Bryozoans. This specimen shows the
white imprint of a Bryozoan called Fenestella.

Conasauga shale

89

Mississippian Shale (CUP) with fenestella fossil

Shale from Cumberland Plateau

Chert

Rock Class: Felsic, Sedimentary
Primary Minerals: microscopic quartz
Secondary Minerals: ----
Province: CUP, V&R, CP

Chert forms in three ways. Bedded chert forms when microscopic creatures called diatoms and radiolarians thrive in warm shallow seas. Their silicon-rich skeletons accumulate on the bottom in layers, usually along with some limestone and mud to form thick beds. Nodular chert often forms in bedded chert and in limestone. This happens when water moving through the chert or limestone bed picks up silicon and redeposits it in openings to form glass-like chert pods or nodules. In both chert forms, the quartz crystals are microscopic in size (called cryptocrystalline quartz), so the chert is silky or glassy in texture, especially in its nodular form.

A final way chert forms is from very fine felsic pyroclastic ash blown from island arc volcanos like those of EBRIP and CT, and deposited on the ocean floor. This "siliceous ooz" is eventually pressed into chert, and finally metachert.

The first photo below is the Fort Payne chert formation, consisting of bedded chert mixed with some limestone and mud. This specimen came from the area where highway 136 crosses Taylor Ridge in the V&R province. Nodules of tan chert found in this formation were prized by American Indians for making arrow points, spear points and blades.

The second photo is a reddish chert nodule from the Ocala limestone formation of the Coastal Plain. This specimen came from the CEMEX limestone quarry in Clinchfield, Georgia.

The final photo is from the CP, and it has the interesting local name of fuller's earth. It is a very silicon-rich clay that overlies the Ocala limestone in places like the Clinchfield CEMEX quarry. In the old days, fuller's earth (in powdered form) was used to clean oil and dirt from newly woven wool fabric. Now, they use soap. This author wonders if, with time, heat and pressure, this fuller's earth (called Twiggs clay in its Georgia formation) would become chert?

Fort Payne bedded chert (V&R)

Chert "nodule" from Ocala limestone (CP)

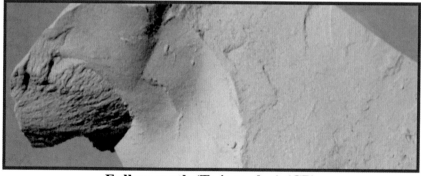

Fullers earth (Twiggs clay) (CP)

Kaolin

Rock Class: Felsic, Sedimentary
Primary Minerals: kaolinite (quartz + aluminum + water)
Secondary Minerals: ----
Province: CP

Kaolin is a soft, perfectly white clayey rock that you can crush with your hands. It is made mostly of the mineral kaolinite. In Georgia, it is found mainly along the Fall Line, but here it is covered by as much as 100 feet of sand and soil. Kaolin mining companies remove the sand and soil and mine this very valuable stuff for inclusion in ceramics, paper, paint, and more. Want to see ceramic made from kaolin? Take a look at your toilet---beautiful!

Georgia kaolin formed between 100 and 65 MYA when sediments from the Mountains and Piedmont washed down into coastal plain swamps near the ocean which reached the Fall Line back then. The sediments were full of feldspar, a mineral rich in silicon and aluminum. The swamps got covered up by other sediments and the clayey sediments became kaolin. The photo shows kaolin from the Thiele Kaolin mine near Sandersville.

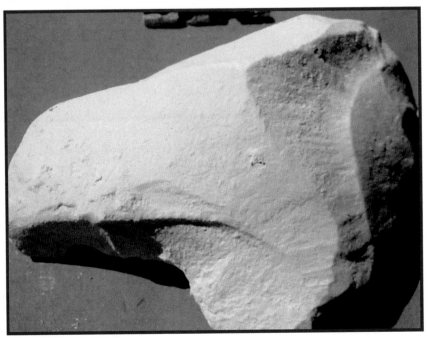

Limestone and Dolomite

Rock Class: Carbonate, Sedimentary
Primary Minerals: calcium carbonate
Secondary Minerals: magnesium-calcium carbonate
Province: CUP, V&R, CP

Limestone can be dark gray, bluish gray, light gray, yellowish or whitish in color. A hand lense may reveal "milky" translucent calcite crystals, and you may see white calcium carbonate veins striping the rock. Sometimes shell creature fossils are visible. Chip off a piece of the rock, grind it to powder and put it in a glass with a little vinegar. If bubbles develop in a half hour or less, then you most likely have limestone (dolomite and marble also fizz).

Limestone forms in warm shallow, (< 200' deep) tropical waters when the calcium-rich bodies of sea creatures like corals, clams and conchs accumulate on the bottom and are pressed into rock by overlying layers. If the calcium-bearing creatures are microscopic in size, then chalk results. Alabama and the White Cliffs of Dover, England have thick chalk deposits, but Georgia has none.

Chert nodules as well as geodes (rounded silicon-rich bodies) develop in limestone. Geodes often have an outer shell of chalcedony (a form of microscopic quartz) with quartz crystals projecting from the shell into a hollow center. They form via silicon-rich water percolating through the limestone and into a rock pocket where the silicon is redeposited in layers and crystals.

Dolomite (usually whitish in color) forms when magnesium replaces some of the calcium in limestone to form calcium-magnesium carbonate. Geologists believe (they don't know for sure) this happens in hot shallow lagoons with few if any currents. Magnesium dissolved in sea water combines with carbon dioxide or with existing limestone, possibly with the help of bacterial mats.

Much more dolomite was formed during the Precambrian and early Cambrian (700-550 MYA) than in later times, including today. This may relate to the fact that creatures that incorporate calcium in their bodies were scarce until Cambrian times.

V&R limestone with white calcium carbonate streaks

Oolitic limestone made of tiny round shell creatures

Ocala limestone from Clinchfield, Ga. with actual (not fossilized) shell pieces.

Dolomite from the V&R

Ga. dolomite can also look grayish

Coal

Rock Class: Sedimentary
Primary Minerals: carbon
Secondary Minerals: ------
Province: CUP

Georgia coal comes from the CUP (Cumberland Plateau). It is black and shinny, probably anthracite (hard coal), though a nickel will scratch it. The coal seams or layers are mixed in with dark shale layers (former mud) that often contain fern fossils.

Georgia coal formed during Pennsylvanian time when the mountains from Africa's crashing were eroding into the Cumberland Plateau foreland basin (present day Lookout Mountian and beyond). Back then, large coastal plain rivers lined with huge swamp forests dumped into the basin. The forests supported tree-size ferns and other fern-like plants whose remains compacted into peat, just like that of the Okefenokee Swamp. Ice age sea level fluctuations repeatedly buried the peat swamps with mud. Over millions of years the peat turned to coal and the mud turned to fossil-rich shale.

Durham Rd. on Lookout Mountain is the former site of the Durham coal mine, a strip mining operation that lasted into the 20th century. Today, the road is lined with piles of fossil-rich waste shale, dug away from the coal seams and piled up in berms. You can dig for fossils there if you contact the Lula Lake Land Trust: http://www.lulalake.org/

Metamorphic Rocks

Quartz Vein Rock
Rock Class: Felsic, Metamorphic
Primary Minerals: quartz
Secondary Minerals: Rarely, iron, copper, rutile, tourmaline, gold
Province: EBRIP, CT, WBR

Quartz vein rock has a glassy appearance. It also occurs as relatively long, six-sided crystals. It is usually transparent or translucent, but can be opaque also. Quartz vein rock ranges in color from colorless to gray, white, yellow, red, violet and even blue. Quartz vein rock is hard (a steel knife won't scratch it).

Quartz vein rock is very abundant in EBRIP and CT. It is much less abundant or absent from other Georiga geologic provinces. This reflects its metamorphic origin.

Quartz vein rock comes from veins formed in other rocks that have lots of quartz in them like granite, sandstone and felsic tuff. During continental collisions, the pushing and shoving opens up tiny cracks in host rock like granite. Water (sometimes hot) migrating through the host rock picks up silicon in solution along with other minerals that may be present, like iron, titanium and gold. The water deposits the quartz in the tiny crack. The crack is further enlarged by more pushing and shoving, only to be filled with more vein quartz. This process can continue until a large vein is formed. The host rock eventually erodes away exposing the harder quartz vein which finally breaks up as quartz "float". Very resistant to weathering, it just stays around and accumulates over the eons in quantities far exceeding its actual occurrence in a given section of host rock.

The fabulous mother lode of the Dahlonega Georgia gold belt was created when both quartz and gold were picked up by water and deposited in large gold-rich quartz veins. It happened like this. As the EBRIP converged on Laurentia, ocean water was drawn deep into mafic magma chambers associated with ocean crust subduction. The water picked up gold and other valuable metals from these deep places. The super heated water then expanded, flushing through porous layers of felsic pyroclastic host rock, leaching sodium and calcium and depositing gold and other metals.

Later, water picked up quartz and gold in solution from the host rock and deposited it in quartz veins. These gold veins were formed during the crashing episodes that metamorphosed EBRIP rocks.

If the cracks where quartz forms are big, large quartz crystals can "grow." If these crystals contain iron as an impurity, they become amethysts. The Jackson Crossroads amethyst mine in Wilkes County produces world-class amethysts. Their web site tells how you can dig there for a fee. Here is the web site (year 2011): http://www.wncrocks.com/resources/Collecting%20site%20jackson%20crossroads.htm

From Web

Quartzite

Rock Class: Felsic, Metamorphic
Primary Minerals: quartz
Secondary Minerals: sometimes hematite iron
Province: EBRIP, CT, WBR (metasandstone)

Quartzite looks like thousands of sand grains mashed together. It often looks slightly more glassy than sandstone, because the grains got hot enough (via metamorphism) to partially melt them together. When you break sandstone, the sand grains separate from each other to form the break. When you break quartzite, it splits right through some of the sand grains because they have been welded together by metamorphism.

Quartzite forms when sandstone is subjected to great heat and pressure over time. Sandstone usually forms from ancient sand beaches, river margins and sandy river deltas that harden into rock. Mammoth Cave, the longest cave in the world, is capped by a thick ceiling of sandstone formed by the Green River long ago.

EPRIP and CT quartzite probably originated as grains of sand-size quartz pyroclastic material deposited (and possibly sorted by water) on the underwater flanks of a volcanic island like EBRIP or CT. Not abundant in EBRIP and CT, quartzite is often associated with quartz muscovite schist, suggesting a progression from quartz-rich, to quartz and feldspar-rich original material.

Granofels

Rock Class: Felsic to Mafic Metamorphic
Primary Minerals: quartz, mafic minerals like calcium plagioclase
Secondary Minerals: Sillimanite
Province: EBRIP, CT

Granofels is a confusing (even the name is weird!) rock that looks a little like quartzite and a little like quartz float. It is sometimes mistaken for both. Granofels is usually dense and heavy, displaying a crystalline appearance. It has a grainy, sugary or silky texture with **little or no internal layering,** and is made mostly of quartz often accompanied by mafic minerals. Granofels often contain enough calcium plagioclase to be called calc-silicate granofels.

Granofels form from other rock types in a manner similar to quartz veins. You could consider them the mafic version of quartz vein rock. In fact, they are most often encountered in association with mafic rock types like amphibolite, and they also accumulate in quantities that exceed their actual occurrence in a given section of rock.

Granofels with tan sillimanite

Metachert

Rock Class: Felsic, Metamorphic
Primary Minerals: quartz
Secondary Minerals:--
Province: EBRIP, CT

Georgia metachert looks white or gray and has a very fine creamy (but not glossy like milky quartz) texture. It is hard (steel knife won't scratch), because it is made mostly of quartz. The quartz crystals composing metachert are microscopically small (cryptocrystalline), producing the creamy texture.

EBRIP and CT, metachert likely formed from extremely fine (dust or smaller) particles of pyroclastic quartz that settled very far out (10 miles) in the ocean from the volcanic island. Silicon from this material is picked up by water and deposited as microcrystalline chert. Metamorphism can't do much with chert since it is just quartz, so metachert looks like chert. Chert can also form in limestone (p. 91).

EBRIP metachert

Slate, Phyllite and Metasiltsone
Rock Class: Felsic, Metamorphic
Primary Minerals: quartz, feldspar, muscovite
Secondary Minerals: ------
Province: WBR, CT

Slate is metamorphosed shale (p. 89). It looks like a harder, more flattened version of shale. If the shale had fossils, they may be obliterated when it is metamorphosed into slate. Slate may have a dull luster. CT has some slate, but it is not abundant

Phyllite is further metamorphosed slate. It looks like slate with a very satiny or shiny luster. WBR has both slate and phyllite along its border with V&R, in the Snowbird Group of rocks.

The WBR experienced light metamorphism on its northwest side, moderate metamorphism in the middle and higher metamorphism on its southeast side, so you get slate in the northwest that grades into phyllite and then schist in southeast WBR. The metamorphic progression for this material is: shale—slate—phyllite—schist.

In the WBR you may encounter a rock type geologists call metasiltstone or siltstone. This rock is very abundant in the Snowbird Group of rocks north and east of Gatlinburg, Tennessee where it is called the Pigeon siltstone. Metasiltstone is very fine-grained like slate, but it doesn't break into thin sheets like slate. It is thicker-bedded. Pigeon siltstone is gray-green to brown in color and is usually laminated (marked with thin lines).

WBR slate

WBR Phyllite

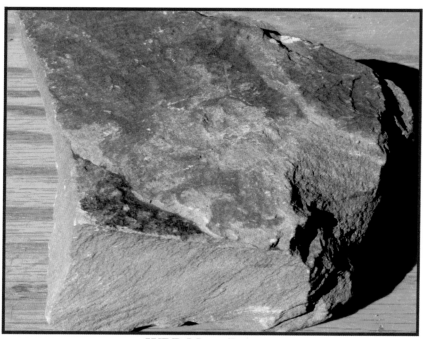

WBR Metasiltstone

Quartz Muscovite Schist

Rock Class: Felsic, Metamorphic
Primary Minerals: quartz, muscovite
Secondary Minerals: sericite, sillimanite, biotite
Province: EBRIP, CT, WBR

What is schist? It is usually a shiny metamorphic rock that contains a lot of muscovite mica, though there is also talc schist (soapstone) and chlorite schist (p. 122). When potassium feldspar (mud) is subjected to the heat and pressure of intense metamorphism in the presence of abundant water, it gets hydrated, and eventually becomes muscovite mica. With continued pressure, the mica grains are flattened and aligned parallel to each other, giving them a very slick and shiny appearance we call schistose.

Quartz muscovite schist is a mixture of quartz grains and muscovite grains (occasionally biotite) that can take on a variety of configurations based on the ratio of quartz to muscovite as well as the kind and degree of metamorphism the original material experienced. It often appears as a jumble of quartz and muscovite grains mixed up together. In some specimens, you may be able to see pea-size rounded quartz grains, probable remnants of pyroclastic quartz fragments. More metamorphosed specimens display layers or bands of quartz separated by thin, very shiny and schisted layers of muscovite. In still others, the quartz is mixed up with very fine grained mica (called sericite) that looks like white or silvery powder. Your fingernail can scratch the sericite which is a common ingredient in auto paint and face makeup.

Specimens of this rock type that were subjected to both hydrothermal alteration and intense metamorphic deformation may contain a buff or tan-colored mineral called sillimanite that forms a fibrous looking matrix called fibrolite between the quartz layers. The fibrolite causes the rock to be very foliated, breaking into thin sheets. Pyrite, imparted during hydrothermal alteration stains the rock a reddish purple color, and the muscovite makes the whole thing very shiny. What a glamorous rock! It is actually very distinctive and can be, "traced for miles along the strike."

EBRIP and CT quartz muscovite schist most likely began life as felsic pyroclastic rock blown from an island arc volcano. It settled on the under water volcano flanks as felsic pyroclastic tuff. The tuff was a jumble of quartz and potassium feldspar pieces from pea-size down to tiny glass shards. In many places, hot water from volcanic vents (black smokers) flushed through the tuff. Sodium and calcium were leached out and iron pyrite was deposited, imparting a distinctive reddish purple color to the tuff. In some cases gold and other metals were deposited along with the pyrite, providing host rock from which metal-laden quartz veins formed.

Later, when proto-Africa crashed into Laurentia, the leached felsic tuff was metamorphosed into the quartz muscovite schist. Heat, pressure and time squeezed water into the feldspar turning it into muscovite. Continued squeezing aligned all the muscovite grains in the same direction and flattened them. The quartz in the felsic tuff was forced into veins, both thick and thin that alternate with thin layers of aligned muscovite.

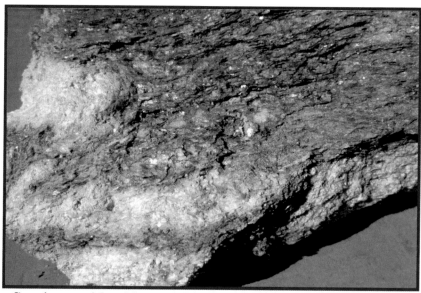

Specimen with white sericite and reddish purple pyrite, sure signs of hydrothermal alteration

Specimen showing shiny schisted appearance and purple from pyrite imparted during hydrothermal alteration.

Specimen made mostly of quartz with very thin muscovite layers. The top and bottom of this specimen are shiny with muscovite. This version is found near the Brevard Fault Zone.

Muscovite Schist
Rock Class: Felsic, Metamorphic
Primary Minerals: muscovite
Secondary Minerals: quartz, garnet
Province: EBRIP, CT, WBR

Muscovite schist (also called mica schist) is similar to, and grades into quartz muscovite schist. It has more muscovite than quartz. It usually contains mostly small muscovite mica flakes mixed in with a little quartz, but near the Brevard Fault zone (Chattahoochee River follows this fault) the mica flakes are large flattened and greasy looking, and quartz is virtually absent. The mica flakes are very shiny, and sometimes dotted with small garnet crystals. They slough off and wash into the creeks as rounded button schist or schist pennies. The lateral sliding movement that created the Brevard Fault (just a big stress crack) produced this effect.

EBRIP and CT muscovite schist is a metamorphic rock likely created when fine mud and ash, rich in silicon, aluminum and potassium was blasted from an island arc volcano, and settled far out in the ocean around the volcanic island. It settled away from the hot water vents and was, for the most part, not leached like the quartz muscovite schist. It was metamorphosed into schist much later when proto-Africa hit.

When EBRIP slid up on Lurentia, it shoved up a bunch of mud sediments ahead of it. This mud could also be the source of some of the EBRIP muscovite schist.

WBR muscovite schist most likely began life as mud washed into the WBR East Graben when Rodinia rifted apart (page 12). Shale formed from the mud. Progressive metamorphic deformation turned the shale into slate, then phyllite and finally into muscovite schist, the end result of intense metamorphosm of felsic mud sediments.

Muscovite schist showing shiny flattened muscovite crystals. This specimen is from the Brevard fault zone.

Muscovite schist "buttons" that slough off the rock and fill adjacent creeks with "scchist pennies". Tiny garnets (red circle) form in the schist.

Metadacite

Rock Class: Felsic, Metamorphic, Extrusive
Primary Minerals: quartz, plagioclase feldspar (albite)
Secondary Minerals: biotite
Province: CT

Dacite is pronounced, "day site".... rather tricky, eh?
Metadacite is metamorphosed dacite lava. Only rhyolite lava is
more felsic than dacite. Andesite lava is a little more mafic than
dacite. This rock is composed of pea-size crystals of quartz
(sometimes blue-gray in color) and plagioclase feldspar in a fine-
grained matrix of the same minerals.

Dacite lava occurs in subduction-generated volcanic islands like
EBRIP and CT due to fractional melting. During fractional
melting, felsic minerals like silicon, potassium and aluminum melt
first, separating from the subducting ocean crust to form a light
weight magma that "floats" to the top of the volcano, plugging its
opening as a lava dome.

This all happens slowly over time, about as fast as fingernails
grow, and consequently there is time for crystals of quartz and
feldspar to form as the stuff slowly cools. The biggest crystals
form in the hot center of the mix because it cools more slowly than
the magma edges. The quartz and feldspar crystals are called
phenocrysts, and they are embedded in a fine-grained matrix of the
same minerals. Most of this felsic material blows out of the
volcano as pyroclastic rock, but some can bulge up to form dacite
domes, or cough out in blobs of dacite lava. Silicon-oxygen bonds
are so strong, felsic lavas like rhyolite and dacite don't flow well.

Metadacite is abundant in the Carolina Terrane around Lincolnton,
Georgia, but it is rare or absent in the EBRIP according to the
USGS Geologic map. One possible explanation for this difference
relates to metamorphism. In the Carolina Terrane, as metamorphic
grade increases, the metadacite begins to look a lot like granite
gneiss. EBRIP has experienced significantly higher levels of
metamorphism than the Carolina Terrane, so its metadacite may
now be in the form of granite gneiss instead of metadacite.

Lincolnton metadacite with feldspar phenocrysts

Specimen showing quartz phenocrysts

Granite Gneiss

<u>Rock Class:</u> Felsic, Metamorphic, Intrusive
<u>Primary Minerals</u>: quartz, potassium feldspar, plagioclase feldspar,
<u>Secondary Minerals:</u> biotite, muscovite
<u>Province:</u> EBRIP, CT, WBR

Granite gneiss is granite that has been deformed by the heat and pressure of <u>metamorphism</u>. This process sometimes causes the light-colored minerals like quartz and potassium feldspar to migrate into swirling bands in the rock. Fresh (unweathered) granite gneiss from gravel quarries looks very different from granite gneiss in old rock outcrops that have experienced millions of years of weathering. Unweathered granite gneiss looks firm and gray like granite, but it usually displays the swirling bands that mark it as granite <u>gneiss</u>. Weathered granite gneiss can be crumbly and "rotten" looking. The bands may be very hard to see or absent.

Granite gneiss is much more abundant than granite in the EBRIP and CT. Examples of regional granite gneiss bodies are the Athens granite gneiss, Lithonia granite gneiss, Jacksons Crossroads granite gneiss and Woodville granite gneiss.

When an island arc like EBRIP or CT crashes into the mainland, its weight pushes the underlying continent crust down toward the mantle. As a result, the crust melts and is emplaced in the island arc above as granite pultons. Later continental crashing (Africa) metamorphoses the pultons into granite gneiss.

A walk along the shores of Lake Hartwell will turn up an interesting form of granite gneiss called <u>megacrystic microcline gneiss</u>. Large crystals of microcline (potassium feldspar) are embedded in a matrix of quartz, plagioclase feldspar and biotite. Geologists generally don't know why this unique gneiss formed.

The WBR has just a couple of granite gneiss bodies. Located in Pickens County, these plutons are actually hunks of 1 billion-year-old Grenville basement gneiss exposed by erosion of overlying Snowbird Group rocks. Geologists call this type of exposure a geologic window. This very old granite gneiss has elongate eye-shaped crystals of quartz or feldspar, the result of intense metamorphism. Augen is a German word that means eye-shaped.

Unweathered quarry specimen showing metamorphic banding

Weathered specimen with metamorphic banding

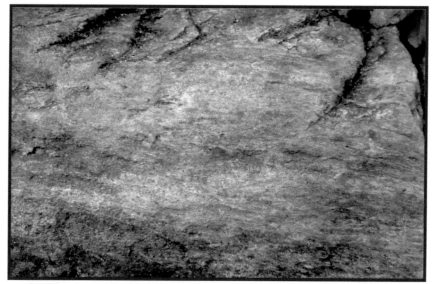

WBR Grenville basement augen granite gneiss with pink potassium feldspar augens (means eye-shaped)

Granite gneiss pegmatite (meaning big crystals) with gray quartz and white feldspar phenocrysts. Pegmatites often form around the edges of granite and granite gneiss plutons.

Biotite Hornblende Gneiss

Rock Class: Intermediate, Metamorphic, Intrusive
Primary Minerals: quartz, biotite, potassium feldspar, plagioclase
feldspar (sodium-rich), hornblende
Secondary Minerals: garnet
Province: EBRIP, CT

Biotite hornblende gneiss is grayish or black in overall appearance.
A closer look reveals a "salt and pepper" mixing of light and dark
colored minerals. The light colored minerals are quartz, potassium
feldspar and plagioclase feldspar (usually sodium-rich). The dark
minerals are biotite and/or hornblende.

EBRIP and CT biotite hornblende gneiss is a very heterogeneous
rock type, showing variety in mineral composition and crystal size.
Metamorphic banding is often present with quartz and potassium
feldspar oriented in bands, but there is a fine-grained variety found
at Watson Mill Bridge State Park, Thompson Mills Forest
(Braselton) and other locations that strongly foliates into thin slate-
like plates and lacks significant banding.

Some specimens of biotite hornblende gneiss have mostly biotite
with very little hornblende. These we would call biotite gneiss.
The more mafic hornblende may dominate in other specimens, and
we would call them hornblende gneiss.

Biotite and hornblende are very difficult to distinguish from each
other in this rock type. Hornblende is more prone to appear as
needle or rod-shaped crystals in the rock, but this characteristic is
frequently obscure. A properly done scratch test with a nickel can
help.

These rocks usually have some foliation (layering), imparting a
recognizable top, bottom and sides to a given specimen. Choose a
black section of the top or bottom, and begin scratching with a
nickel coin. Be sure you are scratching the black stuff, not
something else! If a black or gray powder quickly emerges with
light scratching, and after blowing it aside, you have a groove that
can't be wiped off; then it's biotite. If the black mineral resists
scratching, producing little or no dust; then it is likely hornblende.
Biotite hornblende gneiss most likely originated as an intermediate

intrusive igneous rock called <u>diorite.</u> When volcanic island arcs like EBRIP and CT are created, fractional melting occurs, producing felsic, intermediate, mafic magma and ultramafic magma. Diorite is the most likely candidate for the intermediate magma. It forms the core of the Andes mountains of South America, a present-day example of ocean crust subducted beneath a land mass. The diorite plutons were metamorphosed into biotite hornblende gneiss when the Carolina Terrane, and then proto-Africa crashed into Laurentia, and exposed eons later via erosion.

Biotite hornblende gneiss is the most abundant rock in the EBRIP and CT because it was the most abundant magma produced by fractional melting when these volcanic islands formed.

Biotite gneiss with a nickel scratch from the Brevard Fault zone where strike-slip deformation has produced the foliated (flat layered) appearance.

Side view of above specimen

Salt and pepper look of biotite gneiss

Biotite hornblende gneiss with brownish garnet crystals

Amphibolite and Metagabbro

Rock Class: Mafic, Metamorphic, Intrusive or Extrusive
Primary Minerals: hornblende, plagioclase feldspar (calcium-rich),
little or no quartz, no biotite
Secondary Minerals: pyroxene
Province: EBRIP, CT, WBR

Amphibolite looks black, and is usually spotted with roundish clear
to white plagioclase crystals. It is heavy, crystalline, hard to crack,
and it has a rounded weathering pattern. A nickel will not scratch
the black mineral (hornblende) in amphibolite. The transparent-
looking plagioclase feldspar crystals can look like quartz, so check
the weathered edge of the specimen for the whitish or yellowish
appearance of weathered plagioclase as a clue.

Unlike diabase, amphibolite is not found in long narrow dikes.
Instead, it occurs as lenses of rounded rock, mixed in with other
rock types like biotite gneiss and quartz muscovite schist.

EBRIP and CT amphibolite is most likely the metamorphosed
equivalent of mafic magma/lava that intruded into these island arcs
during fractional melting of ocean crust.. Amphibolite could also
be mafic ocean crust squeezed between an island arc and a
continent, and deposited on the surface with ultramafic rocks..

Metagabbro has a mineralogical makeup similar to amphibolite,
but its large mineral grains suggest it formed as a pluton deep
underground where very slow cooling allowed larger crystals to
grow than its above-ground counterpart basalt.

Soils weathered from amphibolite and metagabbro are more fertile
than soils derived from other rock types. This is due to the high
calcium content of the plagioclase feldspar in amphibolite. The
calcium raises soil PH, enhancing its ability to provide the macro-
nutrients plants thrive on. The rich floras associated with these
soils often include rare plants.

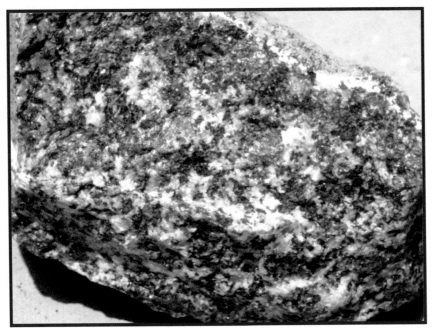

Amphibolite with abundant white calcium plagioclase crystals

Athens, Georgia amphibolite

Metagabbro from near Eatonton, Ga., showing the big crystal size that helps identify it

The rounded weathering pattern of amphibolite

Highly Weathered Mafic Rock

Rock Class: Mafic, Metamorphic
Primary Minerals: Highly weathered mafic minerals
Province: EBRIP, CT

This is almost a catchall category. It refers to highly weathered mafic rock, probably from various sources. It appears more like a body of tan-colored dirt clods than rock. Sometimes this "rock" has brown or black streaks or spots that may be relict pyroxene. Examination with a hand lens sometimes shows abundant tiny pits, possibly places where other mineral grains have been removed by weathering. It is described here because it occurs throughout the EBRIP and CT adjacent to solid rock of presumably similar age.

The absence of quartz in this material indicates it is mafic in origin. Crumbled pieces of the stuff fizz furiously in hydrogen peroxide (a test for iron), further suggesting its mafic character.

The fact that it is more highly weathered than surrounding rock also suggests a mafic origin, since mafic rocks weather more rapidly than felsic and intermediate rocks, due to high iron (a highly erodible mineral) content.

A possible source for this material is mafic pyroclastic rock blown from mafic magma sources when EBRIP and CT formed.

Specimen with dark mafic mineral, possibly pyroxene

121

Serpentinite, Chlorite schist, Talc schist (Soapstone)
Rock Class: Ultramafic, Metamorphic, Extrusive
Primary Minerals: olivine, pyroxene
Secondary Minerals:----
Province: EBRIP, CT, WBR

The bottom "half" of ocean crust is composed of an ultramafic igneous rock called peridotite, which is very rich in the minerals olivine and pyroxene. Dunite, one of several varieties of peridotite is composed of 90% olivine, and is part of this lower ocean crust mix.

During island arc formation, ocean crust is subducted beneath the developing island. Some of the peridotite in ocean crust gets subjected to hydrothermal alteration, and is converted to talc schist (soapstone), a soft white metamorphic rock you can scratch with your fingernail.

Eventually, the island arc crashes into the mainland, trapping a bunch of the ocean crust (peridotite and talc schist) in the fault zone. Peridotite is very unstable on the surface. It quickly combines with water in a process called retrograde metamorphism. The olivine is converted to serpentine, changing the rock from peridotite (igneous) to serpentinite (metamorphic), a greenish gray rock that often has flaky green crystals that look like serpent scales!

Drive along highway 76 in the Georgia Mountains to the Popcorn Overlook, a few miles west of highway 76's intersection with 197. You will be right on the Hayesville fault where ocean crust got trapped between Laurentia and EBRIP. Here, you will find outcrops of serpentinite and talc schist. This formation contains veins of the mineral asbestos that formed in the serpentinite just like quartz veins form. The asbestos looks like rotten wood.

Remember the idea of fractional melting during volcanic island formation (page 39)? Geologists now know that an ultramafic magma can form during fractional melting and be emplaced in the developing island arc along with the felsic, intermediate and mafic magmas produced. This ultramafic magma is made mostly of pyroxenite, a rock containing the ultramafic mineral pyroxene.

This happened in the development of both EBRIP and CT. When these island arcs crashed into the mainland, some of the pyroxenite was metamorphosed into chlorite schist, a very soft green rock similar to talc schist. Further crashing caused the soft chlorite schist and pyroxentie to be pushed up to the surface and squeezed out across the land in a thin sheet called an allochthon. The ultramafic rocks of Soapstone Ridge in Dekalb County, those on the western shore of Lake Richard B. Russell and the ultramafic "ring" around Brasstown Bald are most likely remnants of ultramafic pyroxenite allochthons.

Serpentinite from Popcorn Overlook

Asbestos forms veins in the serpentinite

Serpentinite above with a layer of soapstone below

Probable metapyroxenite from Russell Lake allochthon

Marble
Rock Class: Carbonate, Metamorphic
Primary Minerals: calcium carbonate
Secondary Minerals: magnesium carbonate
Province: WBR

Marble is white. Crack it and you can see sparkly calcium carbonate crystals. Quartz veins can also form in marble. Powdered marble is much less likely to bubble in vinegar than limestone, but it's worth a try as an identification help.

Georgia marble is found in the WBR province along the Murphy Marble Belt, a strip of marble and related rocks that winds across north central Georgia through the towns of Mineral Bluff, East Elijay, Jasper and Tate.

The marble was once a layer of limestone covering the Great Smoky Group, Dean formation of sediments that were deposited in the easternmost WBR graben when Rodinia rifted (p. 31). When Africa crashed millions of years later, the Dean and Murphy sediments were folded into a huge downward fold called a syncline and the Murphy limestone was metamorphosed into the Murphy marble.

Tate marble from the Murphy marble belt

APPENDIX 1: GEOLOGIC TIME SCALE

PERIOD NAME	APPROXIMATE TIME
Quaternary	1.5 MYA-Present
Tertiary	65-1.5 MYA
Cretaceous	150-65 MYA
Jurassic	200-150 MYA
Triassic	250-200 MYA
Permian	300-250 MYA
Carboniferous (Pennsylvanian Mississippian)	370-300 MYA
Devonian	415-370 MYA
Silurian	445-415 MYA
Ordovician	490-445 MYA
Cambrian	550-490 MYA
Precambrian	Before 550 MYA

Alphabetical Index

Made in the USA
Middletown, DE
14 October 2020

21954358R00075